Electric Vehicle Fluid Testing

This book presents the latest advancements, innovations, and discoveries in the testing and analysis of fluids for electric vehicle (EV) tribological applications. It covers all major aspects of fluid testing and evaluation, providing readers with essential information for enhancing the efficiency and durability of EV drivelines. Combining fundamental research with practical insights from leading global experts, this book:

- Presents cutting-edge testing methodologies for EV driveline lubricants and coolants.
- Discusses the latest developments in electrified tribotesting for EV lubricants.
- Features real-world data from academic and industrial testing case studies.
- Includes previously unpublished results from electrified tribotesting and other physicochemical analyses of EV fluids conducted in academic and industrial settings.
- Offers a compendium of critical issues related to fluid-induced failures in EVs.

By uniting contributions from pioneering industries and academic researchers, this book is an invaluable reference for engineers and scientists in tribology, lubrication, and automotive engineering, driving progress in the rapidly evolving EV field.

Electric Vehicle Fluid Testing

Edited by
Leonardo Israel Farfan Cabrera,
Peter Lee, and Ali Erdemir

CRC Press
Taylor & Francis Group
Boca Raton London New York

CRC Press is an imprint of the
Taylor & Francis Group, an informa business

Designed cover image: © 2026 Shutterstock

First edition published 2026
by CRC Press
2385 NW Executive Center Drive, Suite 320, Boca Raton FL 33431

and by CRC Press
4 Park Square, Milton Park, Abingdon, Oxon, OX14 4RN

CRC Press is an imprint of Taylor & Francis Group, LLC

ISBN: 978-1-032-80213-8 (hbk)
ISBN: 978-1-032-80246-6 (pbk)
ISBN: 978-1-003-49606-9 (ebk)

DOI: 10.1201/9781003496069

Typeset in Times
by codeMantra

Contents

Preface

As the world shifts toward more sustainable and efficient transportation, electric vehicles (EVs) stand at the forefront of this transformation, offering a promising solution to reduce emissions and increase energy efficiency. However, the rapid adoption of EVs introduces a range of new challenges in their core systems' design, performance, and durability. Among these, the lubrication and cooling systems that maintain the efficiency and longevity of electric drive units are crucial. These systems face unique issues, such as foaming, aeration, degradation, corrosion, friction, and wear, which can significantly affect the performance of electric powertrains.

This book tackles these challenges head-on, providing a detailed exploration of the underlying science, testing methodologies, and mitigation strategies that can enhance the reliability and efficiency of lubrication and cooling systems in EVs. By delving into the complex mechanisms behind foaming, aeration, and material degradation, we shed light on the factors that impact the performance of EV drivetrains and offer innovative solutions to address them.

In Chapter 1, we examine the phenomena of foaming and aeration in lubrication and cooling systems, explaining how they disrupt fluid properties such as viscosity and density, leading to reduced efficiency and increased component wear. We also explore cutting-edge solutions, such as fluid dynamic optimization and AI-driven predictive maintenance, to mitigate these issues and enhance operational stability.

As electric drive systems evolve, so must the materials and lubricants that support them. In Chapter 2, we dive into the growing concerns surrounding corrosion and conductive deposits, which can cause long-term damage to key components in electrically charged environments. By understanding how materials interact in these conditions, we identify new approaches to corrosion control and highlight the need for innovative solutions to maintain the integrity of EV powertrains.

Chapter 3 shifts the focus to the electrical properties of EV fluids—an often overlooked but crucial aspect of their performance. We introduce new testing methods that assess the conductivity, dielectric strength, and tribological behavior of oils, greases, and coolants. These insights are essential for fluid selection and material compatibility, setting the stage for more reliable and efficient EV systems.

In Chapter 4, we explore the compatibility of non-metal components, such as seals and elastomers, with lubricants in electric vehicles. These materials play a vital role in ensuring the durability of EV systems, and the chapter examines the challenges posed by aggressive additives and high-stress operating conditions.

Chapter 5 explores the relationship between electrical fields and lubricating properties further, discussing rheological testing methods such as electrorheology. These techniques help us understand how electrical fields influence lubricant behavior, providing valuable insights into how viscosity and other properties change under dynamic conditions.

Lubricants are also evaluated for their tribological performance under electrical conditions in Chapter 6. Here, we present electrified four-ball testing to assess bio-oils' effectiveness as additives in lubricants for electric drivetrains. As a case study, we highlight the potential of biodegradable oils to meet both performance and environmental sustainability goals.

Chapter 7 explores lubricated contacts in inert gas environments, where electrical fields accelerate tribochemical reactions in EV components. It introduces innovative testing methods for assessing friction and wear performance in both electrified and unelectrified conditions, contributing to a deeper understanding of the challenges faced by EV systems.

In Chapter 8, we present the electrochemical impedance method for characterizing fluid properties, a powerful tool for real-time monitoring of fluid behavior. This technique offers valuable insights into EV fluids' degradation processes and performance, helping to identify weaknesses and improve system reliability.

Grease plays a critical role in the longevity and functionality of EV components. In Chapter 9, we focus on the essential properties of grease adhesion and tackiness. We explore the latest testing methods to evaluate these properties, providing a comprehensive review of traditional and new approaches to optimize grease formulations for various applications.

Finally, Chapters 10 and 11 delve into the degradation and water resistance of greases and the impact of electrification on bearing components. These chapters offer a thorough examination of the mechanisms behind grease degradation and provide strategies for selecting the right formulations to ensure the long-term reliability and efficiency of EV systems.

This book is an essential resource for engineers, researchers, and industry professionals involved in developing and optimizing lubrication and cooling systems for electric vehicles. Combining theoretical insights with practical solutions provides a comprehensive guide to addressing the complex tribological and electrical challenges faced by the electric vehicle industry. Grounded in experimental data and real-world case studies, this work offers a rich understanding of the materials, testing methods, and design strategies necessary to meet the evolving demands of electric vehicles. With a focus on sustainability, reliability, and performance, this book aims to foster innovation and contribute to the ongoing success of electric vehicle technology.

Editors

Leonardo Israel Farfan Cabrera is an Assistant Professor at Tecnologico de Monterrey. He earned bachelor's, master's, and doctoral degrees in mechanical engineering at Instituto Politecnico Nacional (IPN) in Mexico City. He has been a visiting researcher-professor at the Technical University of Munich, the University of Sheffield, Texas A&M University, and Universidad Nacional del Sur and a Fulbright postdoc at Southwest Research Institute. He has authored or co-authored more than 80 research articles and holds 2 patents. His research is primarily related to electric vehicle tribology, tribotesting, and bio/nanolubricants.

Peter Lee earned a PhD at the University of Leeds, where he became a Royal Academy of Engineering Research Fellow. He established the Tribology Research and Evaluations Group at Southwest Research Institute, a world-leading tribology lab at the forefront of electrification of tribology rigs for the EV fluids industry, working directly with equipment manufacturers. He is a Chartered Engineer, a Fellow of IMechE, and a Fellow and Executive Director of the Society of Tribologists and Lubrication Engineers (STLE), as well as being instrumental in establishing the STLE EV conference.

Ali Erdemir is a University Distinguished Professor and the Halliburton Chair in the Department of Mechanical Engineering at Texas A&M University. In recognition of his research accomplishments, he has received numerous awards and honors. He has authored or co-authored more than 300 research articles and holds 37 U.S. patents. His current research focuses on the development of new materials, coatings, and lubricants for a broad range of cross-cutting industrial applications, including electric vehicles.

Contributors

Michael Anderson
Falex Corporation
Sugar Grove, Illinois

Berthold Bode
Flucon Fluid Control GmbH
Barbis, Germany

Yan Chen
Department of Mechanical Engineering
Texas A&M University
College Station, Texas

Michael De Bilde
Falex Tribology NV
Rotselaar, Belgium

Dirk Drees
Falex Tribology NV
Rotselaar, Belgium

Ali Erdemir
Mechanical Engineering Department
Texas A&M University
College Station, Texas

Emmanuel P. Georgiou
Department of Aeronautical Sciences
Hellenic Air Force Academy
Acharnes, Greece

Simon Hausner
Flucon Fluid Control GmbH
Barbis, Germany

Leonardo Israel Farfan Cabrera
Tecnologico de Monterrey
School of Engineering and Sciences
Monterrey, Mexico

André Heine
Flucon Fluid Control GmbH
Barbis, Germany

Robert L. Jackson
Department of Mechanical Engineering
Auburn University
Auburn, Alabama

Jack R. Janik
Department of Mechanical Engineering
Auburn University
Auburn, Alabama

Michael M. Khonsari
Department of Mechanical and Industrial Engineering
Louisiana State University
Baton Rouge, Louisiana

Anthony Khoraych
Advanced Test and Automation Inc.
Milton, Ontario, Canada

Lijesh Koottaparambil
Department of Mechanical and Industrial Engineering
Louisiana State University
Baton Rouge, Louisiana

Yungwan Kwak
Afton Chemical Corporation
Richmond, Virginia

Peter Lee
Southwest Research Institute
Tribology Research and Evaluations
San Antonio, Texas

Seungjoo Lee
Mechanical Engineering Department
Texas A&M University
College Station, Texas

Hong Liang
Department of Mechanical Engineering
Texas A&M University
College Station, Texas

Lais M. Lopes
Falex Tribology NV
Rotselaar, Belgium

Gregory Miiller
Savant Group
Midland, Michigan

Kartik Pondicherry
Anton-Paar GmbH
Graz, Austria

Cesar D. Resendiz-Calderon
Tecnologico de Monterrey
School of Engineering and Sciences
Monterrey, Mexico

Carlos C. Rubio-Hernández
Tecnologico de Monterrey
School of Engineering and Sciences
Monterrey, Mexico

Sudip Saha
Department of Mechanical Engineering
Auburn University
Auburn, Alabama

Carlos Sanchez
Southwest Research Institute
Tribology Research and Evaluations
San Antonio, Texas

Maik Schröder
Flucon Fluid Control GmbH
Barbis, Germany

Sean Skowron
Mechanical Engineering Department
Texas A&M University
College Station, Texas

Paul Staudinger
Anton-Paar GmbH
Graz, Austria

Mohsen Tajedini
Department of Mechanical Engineering
Texas A&M University
College Station, Texas

1 Foaming and Aeration in Electric Drive Units

Mechanisms, Impacts, and Mitigation Strategies

Anthony Khoraych

1.1 OBJECTIVES OF THIS BOOK

The primary objectives of this book are to provide a comprehensive understanding of the mechanisms and effects of foaming and aeration in Electric Drive Units (EDUs), and to present state-of-the-art strategies for mitigating these phenomena. The book serves as a technical guide for engineers, researchers, and industry professionals involved in the design, development, and testing of electric vehicle (EV) powertrains. By addressing the unique challenges posed by foaming and aeration, the book aims to inform future innovations in fluid dynamics, system design, and performance optimization. The key objectives are outlined as follows:

1.1.1 Understanding Foaming and Aeration Mechanisms

The first objective is to establish a detailed understanding of the underlying mechanisms that lead to foaming and aeration within lubrication and cooling systems in EDUs. This includes exploring how air becomes entrained in fluids, the conditions that promote foam formation, and the subsequent impact on fluid properties such as viscosity, density, and thermal conductivity. Additionally, the book investigates how these phenomena affect the overall performance and durability of EDUs, highlighting areas of potential failure and degradation.

1.1.2 Assessing the Impact on System Performance and Reliability

A key focus of this book is to assess how foaming and aeration influence system performance and reliability. The objective is to quantify the effects on lubrication efficiency, heat transfer capabilities, and component wear. By linking these effects to real-world case studies and experimental data, the book provides a clear picture of how poorly managed foaming and aeration can compromise the operational stability and lifespan of EV powertrains.

DOI: 10.1201/9781003496069-1

1.1.3 Presenting Mitigation Strategies and Design Considerations

Another critical objective is to present effective mitigation strategies that can be integrated into the design and development of EDUs. This includes discussing design modifications, such as optimized flow paths, air-bleed systems, and degassing chambers, as well as chemical solutions, like advanced fluid formulations and additive technologies. The book aims to equip engineers with a toolkit of design and material selection guidelines that can be used to minimize foaming and aeration from the earliest stages of development.

1.1.4 Highlighting Future Trends and Research Directions

The book concludes by outlining future trends and research directions in the field of foaming and aeration management. Emerging technologies, such as nanotechnology-enhanced fluids, real-time diagnostic tools, and AI-driven predictive maintenance systems, are discussed to provide a forward-looking perspective on how these innovations can be leveraged to solve existing challenges and set new benchmarks for EDU performance and sustainability.

By achieving these objectives, the book seeks to position itself as a definitive reference for those involved in the advancement of electric vehicle powertrain systems, ensuring that the insights and strategies presented can be applied to develop more robust, reliable, and efficient EDUs.

1.2 ELECTRIC DRIVE UNITS (EDUs) AND POWERTRAINS IN EVs

1.2.1 Subsystems Considerations

- **Battery Pack**: Battery packs generate a substantial amount of heat during charging and discharging cycles. Thermal management systems must, therefore, ensure efficient heat removal to prevent overheating and thermal runaway. In liquid-cooled battery systems, any entrainment of air bubbles can lead to localized hot spots due to the reduced coolant contact with the battery cells. This condition could compromise cell longevity and degrade battery efficiency. Designing reservoirs with air separators, employing degassing valves, and utilizing non-aerating pump designs are effective strategies to mitigate these issues (Figure 1.1).
- **Electric Motor and Inverter Cooling**: The electric motor and inverter are critical for propulsion and energy conversion. These components generate significant heat, especially during high-load conditions such as rapid acceleration or regenerative braking. Fluid design for these systems must ensure uniform cooling without turbulence or aeration, which could lead to inefficient cooling and noise. The use of non-foaming lubricants and precise control of fluid velocities through channel design can help in maintaining the desired thermal performance (Figure 1.2).
- **On-Board Charger and DC-DC Converter Cooling**: The on-board charger and DC-DC converter have lower thermal loads compared to the motor and battery pack but are still susceptible to the adverse effects of aeration

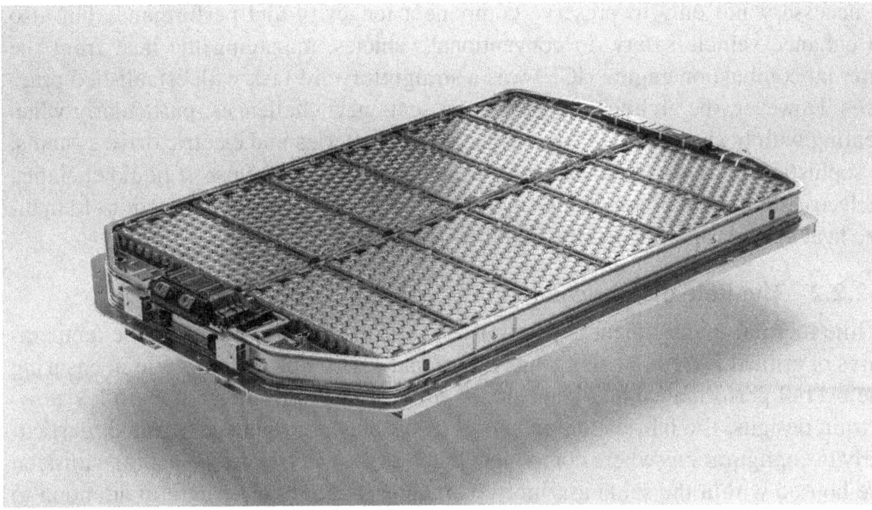

FIGURE 1.1 Battery pack.

FIGURE 1.2 Singular electric motor (left) and motor with inverter assembly on vehicle (right).

and foaming. Designing dedicated cooling loops or integrating these components into the main cooling system requires careful consideration of fluid path design to avoid pressure fluctuations that could introduce air into the system.

1.2.2 Importance of Thermal Management and Lubrication in EDUs

1.2.2.1 Thermal Management in EV Architecture: Design Imperatives and Challenges

The primary objective of the thermal management system in EVs is to maintain temperature stability across the powertrain components, which include the electric motor, battery pack, inverter, and power electronics. Efficient heat dissipation is necessary not only to preserve component longevity and performance but also to enhance vehicle safety. In conventional vehicles, managing the heat from the internal combustion engine (ICE) was a straightforward task with established practices. However, the architecture of EVs presents new challenges, particularly when dealing with heat dissipation from high-voltage batteries and electric drive systems. A sophisticated thermal management strategy can involve the use of liquid coolants, dielectric fluids, and phase change materials, each serving specific functions to manage heat across various operating conditions.

1.2.2.2 The Role of Lubrication in EV Systems

While thermal management is primarily concerned with maintaining the temperatures of critical EV components, lubrication plays a complementary role in ensuring the overall performance and longevity of these systems. In many modern EV powertrain designs, the lubrication and cooling systems are tightly integrated, particularly in configurations where components such as the electric motor and transmission are housed within the same assembly. This integration requires careful attention to how lubrication properties are influenced by aeration and foaming within the shared fluid medium.

Lubrication serves two key purposes in EV systems: reducing friction and wear in mechanical components, and contributing to heat dissipation. Proper lubrication reduces friction between moving parts like bearings, gears, and rotor assemblies, minimizing mechanical energy losses and preventing heat build-up. Additionally, lubricants help transfer excess heat away from these components, supporting the thermal management function.

In systems where lubricants also serve as cooling fluids, ensuring the stability and efficiency of the lubricant is paramount. Any disruption in lubrication properties—such as changes in viscosity or thermal capacity due to air ingestion or foaming—can cause the fluid's heat transfer properties to change from their design values in unpredictable ways and lead to increased mechanical stress on components. This, in turn, can affect the performance and reliability of the EV powertrain.

1.3 UNDERSTANDING BOUND AND UNBOUND AIR IN EV FLUID SYSTEMS AND THEIR SIGNIFICANCE

In the context of EV thermal management and lubrication systems, air can manifest in various forms that significantly affect the fluid's performance. These forms can be broadly classified into *bound air*—which includes foaming, aeration, and dissolved air—and *unbound air*, which is present in the fluid system but not integrated into the fluid itself. Each form of air has distinct characteristics and poses unique challenges to effective fluid management (Figure 1.3).

1.3.1 Bound Air: Foaming, Aeration, and Dissolved Air

Foaming occurs when air is trapped on the surface of a fluid, forming stable bubbles that create a foamy layer. This layer not only disrupts the fluid's thermal transfer properties but can also block or reduce the flow of coolant or lubricant, leading to

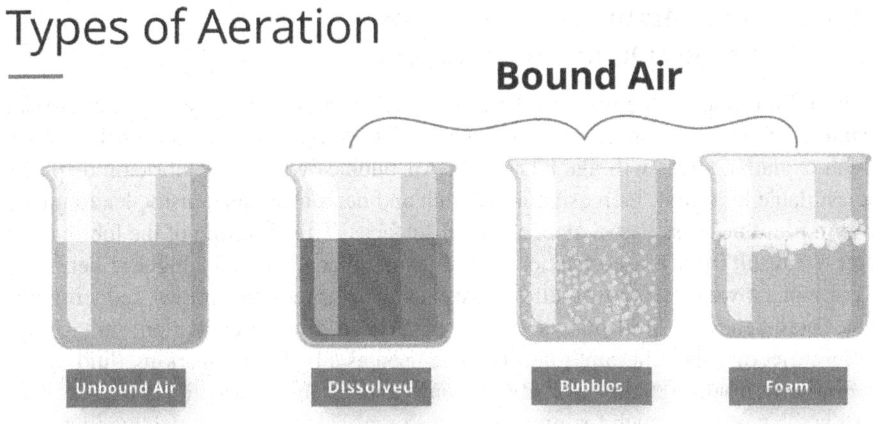

FIGURE 1.3 Types of aeration.

uneven distribution and hotspots. Excessive foaming reduces the system's capacity to transfer heat, increases the risk of overheating, and can impair lubrication, causing metal-to-metal contact and component wear.

Aeration refers to the presence of air bubbles dispersed within the fluid, typically due to turbulence or pressure changes. Unlike foaming, which is surface-bound, aeration bubbles are suspended within the fluid body. Aeration diminishes thermal conductivity and reduces fluid volume, which can cause inconsistent coolant flow rates and lead to cavitation in pumps. Cavitation can damage pump impellers, reduce system efficiency, and contribute to noise and vibration.

Dissolved Air is air that is molecularly dissolved within the fluid, remaining undetectable to the naked eye under normal conditions. While dissolved air may not pose an immediate threat to thermal or lubrication performance, it can become problematic if the system undergoes pressure changes or experiences elevated temperatures. Under these conditions, dissolved air can "undissolve," forming microbubbles that can aggregate into larger bubbles, contributing to both aeration and foaming. This process, known as outgassing, occurs because the solubility of air in fluids decreases as temperature increases or pressure drops. For example, when coolant passes through low-pressure zones in a cooling loop, dissolved air may come out of solution, creating pockets of air that can interfere with flow and heat transfer.

1.3.2 Unbound Air and Its Impact

Unbound air is air that exists outside the fluid as pockets within the system, typically caused by improper filling or system leaks. Although unbound air is not incorporated into the fluid, it can be drawn into the fluid path through turbulence or cavitation, leading to subsequent aeration or foaming. Unbound air also creates the potential for pressure imbalances, which can disrupt the flow dynamics and lead to performance inconsistencies.

1.3.3 Impact of Aeration and Foaming on Lubrication Quality and Performance

Aeration introduces air into the lubricant, which can significantly degrade its performance. Since aeration increases the availability of oxygen, it increases the rate at which oxidative stress will age lubricants. Air bubbles within the lubricant displace the available lubricant, increasing its volume and decreasing its density, leading to a drop in film thickness between mechanical surfaces. This thinning of the lubrication layer can result in increased friction, metal-to-metal contact, and a higher likelihood of component wear. Moreover, air bubbles act as compressible entities, reducing the overall efficiency of fluid pumps and potentially leading to cavitation damage, as well as affecting the lubricant's ability to be used as a hydraulic working fluid.

Similarly, foaming can have severe consequences for lubrication systems. The most notable way in which foaming problems manifest is excessive volume increase, causing fluid to leak from case vents or seals. The presence of foam not only reduces the lubricant's ability to carry heat away from components but also impairs its distribution across surfaces, leaving some areas inadequately lubricated. Excessive foaming can result in localized overheating and increased mechanical wear, especially in high-speed or high-load applications such as electric motors operating under demanding conditions.

1.4 MECHANISMS OF FOAMING AND AERATION IN EDUs

1.4.1 Fluid Dynamics at Gas/Fluid Interfaces

1.4.1.1 Formation of Void Fraction Bubbles and Foam at Gas–Liquid Interfaces

The formation of bubbles and foam in fluid systems is a critical consideration for engineers designing and testing electric drive units (EDUs), as these phenomena can significantly impact both thermal management and lubrication performance. Understanding the principles governing gas solubility, bubble formation, and foaming behavior is essential to predicting and controlling these effects in cooling and lubrication systems.

1.4.1.1.1 Solubility of Gases in Liquids: Henry–Dalton Law

The fundamental relationship describing the solubility of gases in liquids is governed by Henry's Law. Henry's Law states that the concentration of a gas dissolved in a liquid is directly proportional to the partial pressure of the gas above the liquid:

$$C = k_H \cdot P \tag{1.1}$$

where

- C is the concentration of the dissolved gas (mol/L or g/L).
- k_H is Henry's Law constant, which depends on the specific gas–liquid pair and temperature (units: mol/ (L atm)).
- P is the partial pressure of the gas above the liquid (atm).

For multi-component gas systems, Dalton's Law is used in conjunction with Henry's Law to describe the relationship between total pressure and partial pressures:

$$P = P_1 + P_2 + \cdots + P_n \tag{1.2}$$

where P_i is the partial pressure of each gas component. Combining Dalton's Law with Henry's Law gives the concentration of each dissolved gas in the fluid, allowing us to determine the total gas content within the liquid.

The Henry's Law constant (k_H) varies depending on the type of fluid and temperature. For engineers working on EV cooling and lubrication systems, understanding the solubility of gases in these fluids is essential to accurately predict and control aeration, foaming, and bubble formation. Sander's compilation [4] provides an extensive list of Henry's Law constants for water as a solvent.

1.4.1.1.2 Temperature Dependency and Practical Considerations

The k_H values for all fluids decrease with increasing temperature, indicating that gases become less soluble as the fluid heats up. This relationship is crucial for EV applications because, as the coolant or lubricant temperature rises during operation, the potential for gas to come out of solution and form bubbles increases, leading to potential aeration or cavitation issues.

By using appropriate k_H values in conjunction with Henry's Law and Ideal Gas Law calculations, engineers can estimate the amount of gas that may come out of solution under different operating conditions. This data is essential for optimizing fluid system design, selecting appropriate degassing strategies, and ensuring reliable performance in the demanding environment of electric drive units.

1.4.1.2 Void Fraction and Bubble Formation in Liquid Systems

Bubbles form when dissolved gas comes out of solution, a process commonly driven by changes in pressure or temperature. Determining how much gas emerges from the liquid first involves understanding the solubility of the gas in that particular fluid under given operating conditions. A solubility analysis, conducted at equilibrium and over sufficient time, provides an estimate of the quantity of gas that will be released from the liquid phase. This step provides a baseline amount of gas expected to migrate out of solution and potentially form bubbles.

Once the amount of gas to be released is known from the solubility analysis, the Ideal Gas Law can be applied to estimate the volume this released gas will occupy at the specified pressure and temperature conditions. The Ideal Gas Law is given by:

$$V_g = \frac{nRT}{P} \tag{1.3}$$

where

- n is the number of moles of gas.
- R is the ideal gas constant (8.314 J/mol K)).
- T is the absolute temperature (K).
- P is the pressure (Pa).

For a known concentration of dissolved gas C (in mol/L), the number of moles of gas in a given volume V of liquid is $n = C \cdot V$. Substituting n into the Ideal Gas Law gives:

$$V_g = \frac{C \cdot V \cdot RT}{P} \tag{1.4}$$

These two equations provide an estimate of the gas volume released under specific pressure and temperature conditions.

The void fraction uses the estimated gas volume found from the ideal gas law to anticipate bubble formation in cooling or lubrication systems. The void fraction (α) is a dimensionless parameter that represents the volume fraction of gas within a liquid–gas mixture. It is defined as the ratio of the volume of gas to the total volume of the mixture:

$$\alpha = \frac{V_g}{V_g + V_l} \tag{1.5}$$

where

- V_g is the volume of gas within the mixture.
- V is the volume of the liquid.

1.4.1.2.1 Connecting Solubility, Gas Volume, and Void Fraction

- **Solubility Analysis**: Determines how much gas leaves the liquid, providing the amount (moles) of liberated gas.
- **Ideal Gas Law**: Converts the amount of liberated gas into a corresponding gas volume under known temperature and pressure.
- **Void Fraction Calculation**: Uses the estimated gas volume, along with the total mixture volume, to calculate the void fraction (α).

It is important to note that the solubility analysis itself does not directly calculate the void fraction. Rather, it provides the necessary data (i.e., how much gas is released) to then estimate the gas volume using the Ideal Gas Law. With the volume of gas in hand, engineers can determine the void fraction, thereby anticipating bubble formation tendencies, designing appropriate degassing strategies, and ensuring reliable performance in cooling or lubrication systems.

1.4.1.3 Incorporating the Bunsen Coefficient into Void Fraction and Bubble Formation Analysis

The Bunsen coefficient, denoted as B, is another useful parameter for characterizing the solubility of gases in liquids. It defines the volume of gas that can dissolve in a unit volume of liquid at standard conditions and is commonly used to estimate the potential volume of gas that a liquid can hold at different pressures. Mathematically, it is expressed as:

$$\alpha B = \frac{V_{\text{gas}}}{V_{\text{liquid}} \cdot P} \tag{1.6}$$

where

- αB is the Bunsen coefficient (dimensionless, as it represents a ratio).
- V_{gas} is the volume of dissolved gas (in standard conditions).
- V_{liquid} is the volume of the liquid.
- P is the pressure in bars.

When assessing gas solubility in cooling or lubrication fluids, engineers often rely on various solubility coefficients, each defined under specific conditions. Henry's Law constants, for example, are explicitly temperature-dependent: as temperature varies, the equilibrium solubility of a gas in a liquid changes, often significantly. This makes Henry's Law-based analyses well-suited for scenarios where operating temperatures fluctuate.

In contrast, the Bunsen coefficient is defined at a particular reference temperature and pressure—often standard conditions—and represents the volume of gas (at standard conditions) that can be dissolved by a unit volume of liquid at equilibrium and a given partial pressure of the gas. By definition, the Bunsen coefficient does not incorporate temperature changes directly. While this may initially suggest that it remains "unaffected" by temperature or viscosity, this is more a matter of its definition rather than a true insensitivity to changing conditions. In practice, actual gas solubility will still shift as temperature deviates from the reference state at which the Bunsen coefficient was determined.

Due to these limitations, the Bunsen coefficient is best treated as a rough, baseline estimate that applies when conditions are close to the defined reference temperature and pressure. For more accurate engineering predictions—particularly when dealing with variable or elevated temperatures—Henry's Law constants and their temperature-dependent relationships provide a more reliable tool. By selecting the appropriate solubility parameter and understanding its underlying assumptions, engineers can more effectively anticipate gas release and bubble formation in complex fluid systems.

For example, water has a Bunsen coefficient of approximately 0.04, indicating that it can hold an additional volume of air equal to 4% of the liquid volume for every bar above atmospheric pressure. By contrast, mineral oils used in EV lubrication systems can have Bunsen coefficients ranging from 0.06 to 0.09, suggesting a greater capacity to dissolve air. This means that for mineral oils, up to an additional 9% of their volume could be dissolved air for each bar of pressure above atmospheric. As pressure increases, the amount of dissolved gas increases proportionally, making the Bunsen coefficient a straightforward tool to predict void fraction and dissolved air content in the fluid.

1.4.1.4 Dynamic Dissolution and Dissolution of Air: Time Dependency and Nucleation Processes

While the equations derived in previous sections provide a solid foundation for predicting gas solubility and bubble formation at equilibrium, real-world systems rarely operate under these idealized steady-state conditions. In practical scenarios, the dissolution and outgassing of air into and from fluids are time-dependent processes influenced by factors such as the rate of pressure and temperature changes, fluid dynamics, surface area of the air and the fluid interface, and the presence of nucleation agents. This dynamic behavior is governed by mass transfer principles and the kinetics of gas exchange, which are crucial considerations for engineers designing and testing electric drive units (EDUs).

1.4.1.4.1 Governing Law: Diffusion and Mass Transfer Kinetics

The rate at which gas dissolves into or comes out of solution is described by Fick's First Law of Diffusion, which relates the rate of mass transfer to the concentration gradient of the gas in the fluid:

$$J = -D \cdot \frac{\partial C}{\partial x} \tag{1.7}$$

where

- J is the diffusion flux (mol/m^2 s), representing the amount of gas diffusing through a unit area per unit time.
- D is the diffusion coefficient of the gas in the fluid (m^2/s).
- $\frac{\partial C}{\partial x}$ is the concentration gradient of the gas in the fluid (mol/m^3).

The time required for a gas to reach equilibrium concentration within the liquid depends on both the diffusion coefficient and the initial concentration difference. This relationship is captured in Fick's Second Law of Diffusion:

$$\frac{\partial C}{\partial t} = D \cdot \frac{\partial^2 C}{\partial x^2} \tag{1.8}$$

where

- $\frac{\partial C}{\partial t}$ is the rate of change of concentration with respect to time.
- $\frac{\partial^2 C}{\partial x^2}$ is the second derivative of the concentration gradient.

The solution to Fick's Second Law for a transient system, with initial concentration C_0 and boundary concentration C_b, can be used to model how quickly dissolved air in a liquid approaches the equilibrium state. The system typically has a characteristic length scale L, which might represent the distance over which diffusion occurs, such as the thickness of a slab or the radius of a sphere.

The gradients in Fick's Second Law can be estimated as follows:

- Concentration change in time can be approximated as $\frac{\Delta C}{\Delta t}$
- Concentration change in space can be approximated as $\frac{\Delta^2 C}{\Delta x^2}$, where $\Delta x \sim L$

Substituting the spatial and time approximations into Fick's Second Law:

$$\frac{\Delta C}{\Delta t} \sim D \frac{\Delta^2 C}{\Delta x^2} \tag{1.9}$$

Rearrange to solve for Δt (characteristic time):

$$\Delta t \sim \frac{\Delta x^2}{D} \tag{1.10}$$

Substitute $\Delta x \sim L$: By assuming the characteristic length is L, we find that the characteristic time for dissolution or dissolution is proportional to the square of the diffusion length (L) and inversely proportional to the diffusion coefficient (D):

$$t_{\text{equilibrium}} \approx \frac{L^2}{D} \tag{1.11}$$

If the characteristic length L increases (e.g., larger diffusion distances), the time required for diffusion increases quadratically (L^2). If the diffusion coefficient D is smaller (e.g., diffusion is slower), the time required increases inversely with D.

1.4.1.4.2 Nucleation and Gas Bubble Formation: The Role of Nucleation Agents

Dissolution, or the formation of gas bubbles, is not only governed by diffusion but also by nucleation processes. Nucleation is the initial step in phase transformation where dissolved gas molecules aggregate to form a stable bubble nucleus. This process can occur either homogeneously, without any external influence, or heterogeneously, where impurities, surface roughness, or particulate matter in the fluid act as nucleation sites.

- **Homogeneous Nucleation**: Homogeneous nucleation requires a higher degree of supersaturation and is typically not favored in most fluids due to the high energy barrier needed for bubble formation.
- **Heterogeneous Nucleation**: Heterogeneous nucleation is much more common in practical systems, as the presence of nucleation agents (such as particles, surfaces, or fluid impurities) reduces the energy barrier for bubble formation. This means that even small pressure drops or temperature increases can lead to significant dissolution, forming bubbles that contribute to aeration and foaming.

The rate of nucleation can be described by the following equation:

$$J_n = A \cdot \exp\left(\frac{\Delta G}{kT}\right) \tag{1.12}$$

where

- J_n is the nucleation rate (bubbles/m^3 S)
- A is a pre-exponential factor dependent on the fluid properties
- ΔG is the Gibbs free energy change foe forming a critical-sized nucleus (J).
- k is the Boltzmann constant (1.38×10^{-23} J/K).
- T is the absolute temperature (K).

The presence of nucleation agents lowers ΔG, increasing J_n and thus the likelihood of bubble formation even at relatively low levels of supersaturation.

1.4.1.4.3 Real-Life Implications under Transient Conditions

In real-life EV cooling and lubrication systems, transient conditions—such as sudden accelerations, rapid temperature fluctuations, or pump start-stop cycles—can cause rapid changes in pressure and temperature that do not allow the system to reach steady-state equilibrium immediately. Under these conditions, the dissolution and outgassing of air will lag behind the instantaneous changes due to mass transfer kinetics and nucleation rates.

For example, when the pressure in a cooling loop suddenly drops (e.g., due to conditions like cavitation at a pump inlet or a rapid reduction in system pressure), the dissolved air may not immediately come out of solution. Instead, the outgassing process requires time to propagate throughout the system. This delay can result in transient void fractions and bubble formation in certain areas, while other parts of the system remain supersaturated. This non-uniform distribution of gas bubbles can impair coolant flow stability, create localized hotspots, and reduce heat transfer efficiency.

It is important to distinguish between cavitation and air coming out of solution, as these are separate processes with distinct implications. Cavitation occurs when the local pressure drops below the liquid's vapor pressure, leading to the rapid formation and collapse of vapor bubbles. In contrast, air coming out of solution occurs due to a reduction in pressure below the gas's saturation point, leading to the gradual release of dissolved gas into bubbles. Both phenomena can occur during rapid pressure changes, but their mechanisms and resulting challenges differ.

Similarly, in lubrication systems, sudden pressure changes in transmission or gear systems can cause temporary aeration. Here, bubbles form from dissolved gases coming out of solution and may collapse rapidly, leading to micro-cavitation. Over time, this can result in pitting, surface damage, and reduced lubrication effectiveness. These transient effects must be carefully considered during the design and testing of electric drive units to ensure reliable performance under all operating conditions.

1.4.1.5 Effect of Surfactants and Impurities on Aeration and Foaming

The presence of surfactants and impurities can drastically alter the dynamics of bubble formation and foaming. Surfactants, which are compounds that reduce surface tension, stabilize gas bubbles by forming a protective layer around them, preventing them from coalescing and collapsing. This stabilization effect can increase the persistence of foam and prolong the residence time of bubbles in the fluid, which can be detrimental to heat transfer and lubrication properties.

In contrast, solid particles or other contaminants can have both stabilizing and destabilizing effects on foams. Hydrophobic particles can adsorb to bubble surfaces, acting as barriers that stabilize bubbles and increase foam persistence, much like surfactants. Conversely, hydrophilic particles or contaminants that promote liquid drainage can destabilize foams, leading to quicker foam collapse. The interplay between these effects depends on the wettability, size, and concentration of the particles or impurities present. For example, in industrial systems, solid contaminants can sometimes assist in breaking down unwanted foams, while in other cases, they can exacerbate foam stability and prolong its presence.

The effect of surfactants is often quantified using the Gibbs adsorption equation, which relates the change in surface tension (γ) to the surface excess concentration of surfactants (Γ) at the interface:

$$\Gamma = -\frac{1}{RT}\left(\frac{\partial \gamma}{\partial \ln C_s}\right) \tag{1.13}$$

where

- Γ is the surface excess concentration of the surfactant at the interface (mol/m^2),
- R is the universal gas constant,
- T is the temperature,
- γ is the surface tension,
- C_s is the bulk concentration of the surfactant.

By measuring how the surface tension (γ) changes with surfactant concentration (C_s), we can determine how effectively the surfactant reduces the surface tension and how much of it is adsorbed at the interface. The Gibbs adsorption equation quantifies the surface activity of surfactants and can be used to optimize formulations for foam control (e.g., in firefighting foams or beer).

1.4.1.6 Fluid Density and Mixture Behavior

To understand the impact of dissolved and undissolved air on fluid properties, it is necessary to calculate the density of the gas–liquid mixture. The density of a two-phase mixture is given by the density-weighted average:

$$\rho_{\text{mixture}} = \alpha_g \rho_g + (1 - \alpha_g)\rho_l \tag{1.14}$$

where

- ρ_{mixture}: The density of the gas–liquid mixture (kg/m^3),
- α_g: The gas volume fraction (dimensionless, typically between 0 and 1),
- ρ_g: The density of the gas phase (kg/m^3),
- ρ_l: The density of the liquid phase (kg/m^3).

The presence of gas bubbles decreases the mixture density, which alters the flow characteristics, pressure distribution, and heat transfer properties of the fluid. An increase in gas content results in a reduction in thermal conductivity and an increase in fluid compressibility, both of which are undesirable in thermal management systems.

1.4.2 Thermodynamic Aspects

1.4.2.1 Temperature and Fluid Viscosity

Temperature significantly influences the viscosity of fluids, which in turn affects foam formation and stability. As temperature increases, the viscosity of most fluids decreases, making the fluid less resistant to flow and less capable of holding gas bubbles within its structure. This lower viscosity can lead to:

- **Increased Bubble Coalescence**: Lower viscosity reduces the resistance against bubble movement, allowing bubbles to move freely within the fluid and merge with other bubbles, forming larger gas pockets and promoting foam generation.
- **Reduced Foam Stability**: In low-viscosity fluids, the thin liquid films that separate gas bubbles within the foam are less stable and more prone to rupture. This can cause the foam to collapse more easily, leading to a transient foaming effect rather than a stable foam layer.

Conversely, at lower temperatures, higher fluid viscosity can lead to more stable foams because the fluid's higher resistance to flow traps gas bubbles more effectively and prevents their movement and coalescence. This temperature–viscosity relationship must be carefully considered when selecting fluids for systems that experience a wide range of operating temperatures, such as EV cooling circuits that need to perform in both hot and cold climates.

1.4.2.1.1 Temperature and Surface Tension

Surface tension is another critical property influenced by temperature. As temperature increases, surface tension typically decreases, reducing the cohesive forces between liquid molecules at the gas–liquid interface. This lower surface tension affects foam formation in the following ways:

- **Easier Bubble Formation**: With lower surface tension, less energy is required to create new gas–liquid interfaces. This makes it easier for bubbles to form and persist within the fluid, especially in turbulent flow regions or areas with rapid pressure changes.
- **Lower Film Stability**: The thin films that separate bubbles in foam structures become weaker at higher temperatures due to the reduced surface tension. This makes the foam more prone to breakage, leading to less stable foams that collapse more readily.

The temperature dependence of surface tension is typically described by the Eötvös equation:

$$\gamma = \gamma_0(1 - k \cdot (T - T_c)) \tag{1.15}$$

where

- γ is the surface tension at temperature T.
- γ_0 is the surface tension at a reference temperature.
- k is a temperature coefficient specific to the fluid.
- T_c is the critical temperature of the fluid.

Lower surface tension at higher temperatures means that, while foam may form more easily, it will also be less stable and more likely to break down. This transient nature of foam formation at elevated temperatures can complicate the management of foaming in real-world applications.

1.4.2.1.2 Temperature Effects on Surfactant Behavior

Surfactants—compounds that reduce surface tension and stabilize foam—are highly sensitive to temperature changes. As temperature increases, the effectiveness of surfactants can decrease, impacting their ability to stabilize foam. This reduction in effectiveness is due to changes in the surfactant's molecular arrangement at the gas–liquid interface. At high temperatures, surfactant molecules tend to lose their orientation and packing efficiency, resulting in weaker interfacial films.

- **Critical Micelle Concentration (CMC)**: The CMC, which is the concentration at which surfactants start forming micelles, can change with temperature. At higher temperatures, the CMC usually increases, meaning that more surfactant is required to achieve the same level of foam stability as at lower temperatures.
- **Phase Transitions**: Surfactants can undergo phase transitions at elevated temperatures, such as from a liquid-crystalline phase to an isotropic phase. These phase changes can lead to a sudden loss of foam stability, causing foams to collapse unexpectedly.

In cooling and lubrication systems where surfactants are used to control foam, temperature-induced changes in surfactant behavior can lead to inconsistencies in foam management, making it difficult to predict foam stability under varying operating temperatures.

1.4.2.2 Pressure Changes and Bubble Formation

The formation and growth of gas bubbles in a fluid are governed by both thermodynamic and fluid dynamic principles. When the pressure in the fluid drops below the saturation pressure of the dissolved gas, bubbles begin to form. This process is described by the Rayleigh–Plesset equation, which models the dynamics of a single spherical gas bubble in a liquid:

$$R\frac{d^2R}{dt^2}+\frac{3}{2}\left(\frac{dR}{dt}\right)^2=\frac{1}{\rho}\left(P_g-P_\infty-\frac{2\sigma}{R}-\frac{4\mu}{R}\frac{dR}{dt}\right) \tag{1.16}$$

where

- R is the bubble radius (m).
- $\frac{dR}{dt}$ is the rate of change of the bubble radius (m/s).
- P_g is the gas pressure inside the bubble (pa).
- P_∞ is the pressure far from the bubble (pa).

- σ is the surface tension of the liquid (N/m).
- μ is the dynamic viscosity of the liquid (Pa.s).
- ρ is the density of the liquid (kg/m^3).

This equation captures the interactions between pressure, surface tension, and viscosity, determining whether a bubble will grow, shrink, or remain stable. A rapid pressure drop (e.g., through a constriction) can cause P_∞ to decrease, leading to bubble growth. If P_∞ drops below the liquid's vapor pressure, cavitation occurs, forming vapor-filled bubbles that collapse violently, potentially damaging mechanical components.

Here, P_g (gas pressure inside the bubble) can incorporate gas pressure derived from Henry's Law when bubbles are formed by gas coming out of solution. This allows the Rayleigh–Plesset equation to complement solubility models, dynamically describing how bubbles grow once nucleated.

It's important to distinguish between bubbles formed by cavitation (where pressure drops below the vapor pressure, producing vapor bubbles) and those formed by gas coming out of solution (driven by supersaturation). The Rayleigh–Plesset equation is primarily associated with cavitation but can also describe gas bubble dynamics in cases where P_g reflects gas behavior rather than vapor pressure.

1.4.2.2.1 Cavitation and Pressure-Induced Thermodynamic Effects

Cavitation is a particularly problematic effect of rapid pressure changes in EDUs. It occurs when the local pressure of the fluid falls below its vapor pressure, causing the formation of vapor-filled cavities or bubbles. When these bubbles collapse, they generate high-pressure shock waves that can damage nearby surfaces and lead to pitting and erosion of metallic components. Cavitation not only affects the mechanical integrity of the system but also disrupts coolant or lubricant flow, causing a drop in thermal transfer efficiency.

While the vapor pressure of most lubricating oils is extremely low, making pure vapor cavitation less likely, other effects may still create cavitation-like conditions in electric drive units (EDUs). The combination of entrained air, dissolved gases, and high-speed flows in EDUs can still result in cavitation-like phenomena, warranting attention during design and testing.

The thermodynamic behavior of cavitation can be described using the concept of **enthalpy change** during bubble formation and collapse. The energy required to form a vapor bubble is given by:

$$\Delta H = \sigma \cdot A \tag{1.17}$$

where

- ΔH is the change in enthalpy (J)
- σ is the surface tension of the liquid (N/m).
- A is the surface area of the bubble (m^2)

As pressure decreases and vapor bubbles form, the system undergoes a change in enthalpy, affecting the overall thermal state of the fluid. This is particularly relevant in lubricating systems within EDUs, where localized vapor formation can lead to a sudden drop in viscosity and a loss of lubrication performance, resulting in increased friction and wear.

1.4.2.2.2 Pressure–Temperature Coupling: Polytropic Processes

In many EDU applications, pressure changes are not isolated events but are coupled with temperature changes. For example, when fluid is compressed or decompressed, it experiences a change in temperature according to the polytropic relationship:

$$PV^n = \text{constant}$$

where

- P is the pressure (pa).
- V is the volume (m^3).
- n is the polytropic index, which varies depending on the process (isothermal: $n = 1$; adiabatic: $n = \gamma$).

Although temperature is not explicitly represented in this equation, it is inherently connected through the ideal gas law:

$$PV = nRT \tag{1.18}$$

where

$$V = \frac{m}{\rho} \text{ and } n = \frac{m}{MW}$$

Combining the two equations, we can see that temperature changes during compression or expansion because:

$$T\alpha\frac{P}{\rho} \tag{1.19}$$

Here, temperature is directly proportional to pressure and inversely proportional to density. The polytropic constant n determines how the temperature changes with pressure. In adiabatic compression ($n=\gamma$), where no heat is exchanged, a rapid increase in pressure leads to a corresponding increase in temperature. This temperature rise reduces the solubility of gases in the fluid, causing more gas to come out of solution and form bubbles, exacerbating aeration and foaming.

1.4.3 Mechanical Contributions

1.4.3.1 Mechanical Agitation: Air Entrainment and Bubble Formation

As gears rotate and mesh within the gearbox, they create intense agitation in the surrounding fluid. This mechanical agitation, especially at high gear speeds, can introduce air into the fluid, leading to air entrainment and bubble formation. The meshing of gears causes a rapid displacement of the fluid, which, in turn, creates vortices and turbulent flow patterns. The turbulent eddies can trap air from the gearbox atmosphere into the fluid, generating bubbles that are dispersed throughout the lubricant.

The degree of air entrainment is influenced by several factors, including:

- **Gear Speed and Geometry**: Higher rotational speeds and complex gear geometries (e.g., helical or spiral bevel gears) create more intense agitation, increasing the likelihood of air entrainment. Faster-moving gears generate higher shear forces that mix air into the fluid, especially in high-speed electric drive units (EDUs).
- **Fluid Flow Paths**: Sharp corners, sudden changes in flow direction, and constrictions within the gearbox create localized high-velocity regions where the fluid is more prone to mixing with air. These regions can act as nucleation sites for air bubbles, further exacerbating aeration.
- **Fluid Jets**: High-velocity lubricant jets, commonly used in EDUs for directed lubrication or cooling, can create turbulent mixing and introduce air into the fluid. This effect becomes more pronounced when the jet impinges on surfaces or fluid pools, creating splashing and bubble formation.
- **Fluid Surface Interaction**: The interaction between the moving gears and the fluid surface, particularly in partially filled gearboxes, can introduce additional air into the fluid. Splash lubrication systems, which rely on the gears dipping into the lubricant and flinging it around the gearbox, are especially susceptible to air entrainment due to the creation of splashing and splattering effects.
- **Lubricant Interaction with Spinning Motor Rotors**: In wet motor EDUs, the interaction between the lubricant and the spinning rotors of the electric motor can create intense shear forces and vortices that enhance aeration. The high-speed motion of the rotors can entrain air directly into the fluid, further contributing to aeration in these systems.

The result of mechanical agitation is an increase in the number of air bubbles suspended in the fluid, contributing to aeration. If these bubbles are not removed or allowed to escape from the fluid, they can coalesce to form larger bubbles, which may eventually lead to the formation of foam at the surface or within the fluid body.

1.4.3.2 Shear Forces and Foam Formation

In addition to agitation, shear forces within the gearbox play a significant role in foam formation. Shear forces arise when there is a velocity gradient within the fluid, typically caused by the interaction between moving surfaces (e.g., gear teeth) and the lubricant. These forces can have several effects on the fluid and the formation of foam:

- **Shearing of Entrained Air Bubbles**: As the fluid is sheared between gear teeth, entrained air bubbles are broken down into smaller microbubbles. These microbubbles can accumulate and form a dense dispersion of gas within the lubricant. If not stabilized, these microbubbles can coalesce into larger bubbles, which contribute to foaming. Smaller air bubbles will also have a lower terminal velocity due to smaller buoyancy forces, meaning that air release times will increase.

- **Foam Stability and Shear Stress**: High shear forces can temporarily stabilize foam by creating thinner and more stable films around the gas bubbles. High shear forces cause rapid deformation of the liquid film surrounding bubbles, which can lead to an even distribution and orientation of surfactant molecules at the gas–liquid interface. This uniform surfactant coverage reduces surface tension gradients and stabilizes the bubble film, making it more resistant to rupture and thus temporarily enhancing foam stability. This allows the foam to persist longer than it would under static conditions. This is particularly true when the fluid contains surfactants or impurities that reduce surface tension and promote foam formation.
- **Viscoelastic Properties of the Lubricant**: The viscoelastic nature of certain lubricants, especially those with polymer additives, can contribute to foam stability under shear. As the lubricant is sheared, it may become more resistant to bubble rupture, allowing foam to remain intact under high-shear conditions common in gear meshing zones. This is because as the lubricant is sheared, its molecular and additive components redistribute, creating a more uniform layer of surface-active agents around the bubbles. This balanced coverage reduces uneven tension and helps maintain stable bubble films, making them less likely to rupture under mechanical stress.

Within a gearbox, high rotational speeds and elevated torque levels generate intense mechanical agitation and shear stresses on the lubricant. These stresses not only accelerate the formation of new bubbles through rapid pressure fluctuations and turbulence but can also redistribute surface-active species, stabilizing thin liquid films around existing bubbles. As gears mesh and rotate, their combined mechanical actions create localized regions of varying pressure and flow velocity, causing smaller bubbles to remain suspended and preventing them from coalescing into larger, more easily escaped bubbles. Over time, this cyclical shearing and agitation promotes the persistence of fine, stable foam layers, thereby aggravating aeration and foaming issues under demanding operating conditions.

The combination of shear forces and mechanical agitation within the gearbox environment leads to persistent aeration and foaming issues, particularly at high rotational speeds or under high-torque conditions where the lubricant is subjected to intense mechanical stresses.

1.4.4 Chemical Reactions and Additives

The chemical composition of fluids used in electric drive units (EDUs) has a profound effect on their tendency to foam. The base fluid, whether it be water, ethylene glycol, polyalphaolefin (PAO), or a mineral oil, provides the foundational properties like viscosity, surface tension, and thermal conductivity. However, the propensity for foaming is often determined by the specific additives included in the formulation. Additives, including anti-foaming agents, surfactants, detergents, and dispersants, are used to modify the fluid's properties to optimize performance under different conditions. These additives can either reduce or enhance foaming depending on their chemical interactions with the base fluid and dissolved gases.

1.4.4.1 Chemical Composition and Foaming Behavior

The base chemical composition of a fluid influences foaming through its inherent physical properties:

- **Surface Tension**: Fluids with lower surface tension form foam more easily because less energy is required to create new gas–liquid interfaces. Fluids with high surface tension resist the formation of stable bubbles, reducing foam formation.
- **Viscosity**: While higher viscosity fluids can better suspend and stabilize bubbles within the bulk fluid, they also reduce the terminal velocity of these bubbles, slowing their ascent to the fluid/air interface. In a circulatory system, this slower rate of bubble migration may actually diminish the likelihood of stable foam layers forming at the fluid surface. Conversely, low-viscosity fluids allow bubbles to rise and escape more readily but may provide less intrinsic stability to bubble films. As a result, the relationship between viscosity and foaming tendency is not strictly linear or uniform; it depends on the system's operational conditions, the presence of surfactants or other additives, and the fluid's overall rheological characteristics.
- **Polarity and Hydrogen Bonding**: Polar fluids, such as water and ethylene glycol, tend to foam more due to strong hydrogen bonding interactions, which stabilize the films surrounding gas bubbles. Non-polar fluids, like mineral oils and PAO, are generally less prone to foaming because of weaker intermolecular interactions.

1.4.4.1.1 Additives and Their Impact on Foaming Propensity

Various additives are used in cooling and lubrication fluids to enhance properties like anti-wear protection, oxidative stability, and thermal conductivity. However, some of these additives can inadvertently increase the propensity of the fluid to foam due to their impact on surface tension and bubble stability:

- **Surfactants**: Surfactants reduce surface tension and stabilize foam by accumulating at the gas–liquid interface. While surfactants are beneficial in detergents and dispersants to keep particles suspended, they can increase foaming by forming stable films around gas bubbles, making them less likely to collapse.
- **Detergents and Dispersants**: These additives help keep the fluid clean by suspending particles and preventing deposits. However, they also increase the likelihood of foam formation because they behave similarly to surfactants, reducing surface tension and stabilizing foam.
- **Viscosity Modifiers and Polymers**: Viscosity modifiers, such as long-chain polymers, increase the fluid's resistance to flow, which can trap bubbles within the fluid and increase the persistence of foam. Polymers may also contribute to viscoelastic behavior that enhances foam stability under shear conditions.

- **Oxidation Inhibitors**: While antioxidants are essential for extending the life of the lubricant and preventing degradation, some oxidation by-products can increase the fluid's propensity to foam. For example, oxidation can lead to the formation of polar compounds that act as emulsifiers, promoting foam stability.

1.4.4.1.2 Chemical Reactions and Foam Formation

The chemical reactions that occur in fluids under operating conditions can alter the fluid's chemical composition and change its foaming behavior:

- **Oxidative Degradation**: High temperatures and the presence of oxygen can lead to the oxidative degradation of the lubricant. This process creates acidic by-products, alcohols, and other polar compounds that can increase foaming. For instance, mineral oils subjected to high temperatures can oxidize, leading to the formation of carboxylic acids and ketones, which can stabilize foam.
- **Hydrolysis and Saponification**: In the presence of moisture, ester-based synthetic lubricants can undergo hydrolysis, producing acids and alcohols. The presence of metal ions, such as calcium or sodium, can result in saponification reactions, forming soap-like substances that significantly increase foaming propensity.
- **Interaction with Contaminants**: Contaminants like dirt, water, or metal particles can act as nucleation sites for bubble formation. Additionally, contaminants may react with fluid additives, creating emulsions or complex structures that stabilize foam.

Understanding these chemical interactions and their impact on foaming is essential for selecting the right fluid formulation for a given application.

1.4.4.2 Anti-Foaming Agents: Types and Mechanisms

Anti-foaming agents, or defoamers, are additives specifically designed to reduce or eliminate foam formation. They achieve this by either preventing bubble formation or promoting the collapse of existing bubbles. The effectiveness of an anti-foaming agent depends on its compatibility with the base fluid and the specific operating conditions.

1.4.4.2.1 Types of Anti-Foaming Agents

1.4.4.2.1.1 Silicone-Based Defoamers These are the most commonly used anti-foaming agents in cooling and lubrication fluids. Silicone defoamers contain hydrophobic silicon dioxide particles dispersed in silicone oil. They work by spreading rapidly over the foam surface, disrupting the stability of the bubble walls and promoting bubble coalescence and rupture. The low surface tension of silicone oils allows them to spread easily and enter the foam structure, causing it to break down.

- **Mechanism**: Silicone defoamers reduce surface tension, making the bubble walls thinner and less stable. They also create a film on the bubble surface that promotes drainage of the liquid film, causing the bubble to collapse.

- **Chemical Stability**: Silicone-based defoamers are highly stable and effective over a wide range of temperatures and viscosities, making them ideal for EV applications.

1.4.4.2.1.2 Polyether-Based Defoamers Polyether defoamers are water-dispersible additives that work similarly to silicone defoamers but have different spreading and wetting characteristics. They are often used in water-based fluids like coolants to reduce foam without impacting other fluid properties.

- **Mechanism**: Polyether defoamers destabilize foam by reducing the surface viscosity and elasticity of the bubble walls. This leads to faster bubble drainage and rupture.

1.4.4.2.1.3 Mineral Oil-Based Defoamers These defoamers are typically used in oil-based systems. They contain hydrophobic solid particles, such as silica, waxes, or fatty acids, dispersed in mineral oil. They work by altering the surface properties of the bubbles and facilitating bubble coalescence.

- **Mechanism**: The solid particles in mineral oil-based defoamers disrupt the film around the bubbles and promote the merging of smaller bubbles into larger ones, which collapse more easily due to gravitational forces.

1.4.4.2.1.4 Alkyl Acrylate Polymers These are specialty defoamers designed for fluids that contain high levels of surfactants. They interact with the surfactants in the fluid to suppress their foaming tendency by reducing the formation of stable foam structures.

- **Mechanism**: Alkyl acrylate polymers interfere with surfactant micelle formation, reducing the surfactants' ability to stabilize foam. This interaction lowers the foam's elasticity and promotes faster collapse.

1.4.4.2.1.5 Fluorosilicone-Based Fluorosilicone-based defoamers are specialized foam control agents incorporating fluorinated side groups into the silicone backbone. This structural modification yields exceptionally low surface energy, chemical inertness, and resistance to harsh operating conditions. As a result, these defoamers are particularly effective and durable in demanding environments, such as high-temperature, chemically aggressive, or tightly regulated fluid systems.

- **Mechanism**: Fluorosilicone-based defoamers work by disrupting the stability of foam lamellae—the thin liquid films enclosing individual bubbles. Their low surface tension and strong spreading capability allow them to infiltrate the bubble walls, causing them to thin and eventually rupture. This foam-breaking action is maintained even under severe conditions, ensuring consistent foam control and improved system performance.
- **Chemical Stability**: Due to the presence of fluorinated substituents on the silicone backbone, fluorosilicones exhibit exceptional resistance to chemical degradation, oxidation, and thermal stress. They remain stable and effective

when exposed to high temperatures, aggressive solvents, and corrosive substances, maintaining their defoaming properties longer than conventional silicone-based additives. This chemical stability not only enhances performance but also reduces the need for frequent additive replacement, lowering maintenance costs and improving system reliability.

1.4.4.2.2 Mechanisms of Action for Anti-Foaming Agents

The mechanisms by which anti-foaming agents reduce foam can be categorized as:

1. **Film Drainage Acceleration**: Anti-foaming agents reduce the viscosity of the liquid film surrounding gas bubbles, accelerating film drainage. As the liquid drains, the film becomes thinner and eventually ruptures, causing the bubble to collapse.
2. **Bubble Coalescence Promotion**: Defoamers reduce the surface tension difference between the fluid and the gas–liquid interface, causing bubbles to merge. Larger bubbles are less stable and more prone to collapse due to the increased pressure difference across the bubble wall.
3. **Spreading and Displacement**: Anti-foaming agents like silicone oils spread rapidly across the bubble surface, displacing the fluid film and breaking the bubble structure.
4. **Reduction of Surface Elasticity**: By reducing the surface elasticity of the bubble walls, defoamers cause the bubbles to become less stable and more prone to collapse under shear or compressive forces.

1.4.4.2.3 Relationship between Defoamers and Entrained Air

While defoamers are highly effective at reducing surface foam, their use can sometimes exacerbate entrained air content within the fluid. Some defoamers operate by reducing the size of bubbles, which lowers their terminal velocity and increases air release times. Additionally, defoamers promote the collapse of larger bubbles near the surface, which can result in smaller bubbles being forced deeper into the fluid. This increases the concentration of entrained air and may lead to other issues, such as reduced fluid density, impaired lubrication, or noise in sensitive systems like electric drive units (EDUs).

To mitigate this, specialized anti-entrainment additives or combined formulations are often used alongside defoamers. These additives work by:

- Reducing bubble stability at lower depths to allow entrained air to rise and escape more efficiently.
- Modifying fluid rheology to minimize air retention.

Specialized *air release additives* or formulations that balance both defoaming and air release properties are often used. These include:

1. **Deaerators (Air Release Agents)**: Deaerators promote the coalescence of microbubbles into larger bubbles, which escape more readily from the fluid. These additives are particularly useful in systems with high agitation or shearing, such as gearboxes or cooling loops in EDUs.

2. **Silicone and Non-Silicone Air Release Additives**: These additives destabilize small entrained bubbles without excessively suppressing surface foam, making them a good complement to traditional defoamers.
3. **Anti-Aeration Polymers**: These modify fluid rheology to reduce the retention of entrained air, enabling bubbles to rise and escape more efficiently.
4. **Combination Additives**: Many modern formulations incorporate both defoaming and air release properties, achieving a balance between surface foam suppression and entrained air reduction.

Proper selection of defoamers and anti-entrainment agents is critical to balance surface foam control and the reduction of entrained air content, particularly in high-performance systems like EDUs, where aeration can impact lubrication and cooling efficiency.

1.5 IMPACT OF FOAMING AND AERATION ON EDU PERFORMANCE

1.5.1 Lubrication and Heat Transfer Efficiency

The presence of foam and aeration in electric drive unit (EDU) systems can significantly influence lubrication and heat transfer efficiency, leading to detrimental impacts on overall vehicle performance and component longevity. These phenomena are critical for lubrication film stability and effective thermal management in electric vehicles.

1.5.1.1 Detrimental Effects of Foam on Lubrication Film Thickness

The entrainment of air and formation of foam within lubricants disrupt the formation of a stable lubrication film between moving parts, such as gears, bearings, and other rotating components in EDUs. This disruption impacts lubrication film thickness in the following ways:

1. **Reduction in Film Thickness**: An investigation documented by Li et al. (2023) examines how entrained air bubbles within lubricating oil influence lubrication film thickness in thrust bearings. The presence of air bubbles alters the thermal and flow properties of the lubricant, with thermal effects playing a dominant role. Increased temperatures due to entrained air lower the viscosity of the lubricant, resulting in a thinner lubrication film, while interfacial effects from air-oil interactions further modify pressure distribution within the film. These combined effects lead to reduced film thickness, which can compromise bearing performance, emphasizing the need to minimize air entrainment in lubrication systems.
2. **Increased Friction and Wear**: Prolić and Lepušić (2012) examined how foam formation in lubricating oils impacts their anti-wear performance. Foaming leads to discontinuous lubrication, resulting in increased friction and wear of machine components. The research highlights that foam can cause mechanical damage and cavitation, reducing the lifespan of equipment and escalating maintenance costs. By analyzing various lubricating

oils, the authors assess the dynamics of foam creation and dissipation, as well as the tendency for adhesive-sliding wear at different speeds. The findings underscore the importance of controlling foaming in lubricating systems to maintain effective lubrication and minimize wear.

3. **Impact of Foam Stability on Lubrication Efficiency**: Ren et al. (2023) explored the impact of foam stability on lubrication efficiency and highlight the role of defoaming agents in mitigating foam-related issues. Foam formation disrupts the continuity of the oil film, increasing friction and wear between moving parts and reducing cooling efficiency due to the impaired heat dissipation caused by entrained air. The study emphasizes that even small concentrations of additives, such as 0.005% by weight, can significantly influence foaming behavior. If the solubility of a defoaming agent in lubricating oil is too high, it dissolves into the bulk lubricant, lowering the system's overall surface tension. This reduced surface tension facilitates bubble formation and stabilization, exacerbating foaming rather than preventing it. Conversely, defoaming agents with low solubility concentrate at the oil-air interface, effectively breaking down foam with minimal dosage. The research underscores the importance of carefully formulating defoaming agents to ensure they reduce foam stability while maintaining lubrication and cooling performance, ultimately protecting critical machinery from friction, wear, and overheating.

1.5.1.2 Impact on Heat Dissipation and Thermal Management

Foaming and aeration within the cooling or lubrication system also affect the thermal management capabilities of the fluid. When air becomes entrained within the lubricant or coolant, it decreases the fluid's overall thermal conductivity, leading to inefficient heat dissipation. Key impacts include:

1. **Reduced Heat Transfer Coefficient**: As highlighted by Baran (2007), foaming reduces the effective thermal conductivity of the fluid, which in turn decreases the heat transfer coefficient. This leads to higher temperatures in critical components such as the electric motor and power electronics, ultimately reducing system performance and increasing the risk of thermal degradation.
2. **Localized Overheating**: Manz and Cheng (2006) observed that areas with high concentrations of foam experienced localized temperature spikes of up to 15°C higher than non-foamed regions, causing thermal stress and potential damage to components. The presence of foam acts as an insulating layer, preventing the efficient transfer of heat away from hot spots.
3. **Increased Fluid Degradation**: The article "Foaming in Lubricants" published by STLE in *Tribology & Lubrication Technology (TLT)* highlights the detrimental effects of foaming on lubricant performance and fluid degradation. Foam introduces air into the lubricant, accelerating oxidation, which depletes critical additives and leads to the formation of harmful by-products like sludge and varnish. Additionally, foam impairs heat dissipation, causing localized overheating that further accelerates chemical breakdown and

thermal degradation. These effects compromise the lubricant's ability to provide adequate film formation, increasing friction, wear, and the risk of mechanical failure. The article emphasizes the importance of anti-foam additives, effective system design, and routine maintenance to mitigate foaming, extend lubricant service life, and maintain reliable equipment performance.

1.5.2 Wear and Component Durability

The entrainment of air and formation of foam within lubricants have a direct impact on the wear characteristics of components in EDUs. Inadequate lubrication and increased presence of corrosive elements due to aeration can accelerate wear and tear of critical components, reducing their service life.

1.5.2.1 Increased Wear due to Inadequate Lubrication

1. **Surface Pitting and Fatigue**: When aeration disrupts the lubrication film, surface pitting and fatigue failures are more likely to occur. The article "Preventing Micropitting and Surface Fatigue" from *Machinery Lubrication* highlights how inadequate lubrication leads to micropitting, a form of surface fatigue caused by insufficient lubricant film thickness under high-load conditions. This allows asperity contact between surfaces, resulting in localized stress, the formation of micro-cracks, and eventual material loss through pitting. Proper lubrication with appropriate viscosity and cleanliness is crucial to prevent micropitting, as contaminants like air, water, and dirt can further degrade film integrity and accelerate wear. The article stresses proactive maintenance, including monitoring oil quality and inspecting components for early signs of damage, to mitigate surface fatigue and extend machinery lifespan.
2. **Reduced Load-Carrying Capacity**: The combined findings from two studies performed by Goodwin et al. (2007) and Chun emphasize how aeration reduces the load-carrying capacity of journal bearings. Goodwin et al.'s study on hydrodynamic journal bearings shows that air bubbles in the lubricating oil reduce the load-carrying capacity by increasing oil compressibility and decreasing its effective viscosity, particularly at higher speeds. This reduction compromises the hydrodynamic film's ability to support applied loads. Chun's study on turbocharger journal bearings corroborates this finding, showing that aeration disrupts the lubricant film and reduces its ability to carry dynamic loads effectively. The reduced load-carrying capacity highlights the adverse effects of air entrainment on bearing performance, emphasizing the importance of minimizing aeration to ensure reliable and efficient operation.

1.5.2.2 Corrosive Effects of Aerated Fluids

1. **Oxidation and Corrosion**: Air entrainment accelerates the oxidation process within the lubricant, forming acidic by-products that can corrode metal surfaces. This effect was observed by Obiols (2003), who discusses

the formation of acidic by-products during lubricant oxidation. It highlights that as lubricants oxidize, carbonyl compounds form, leading to increased acidity. The study establishes a correlation between the Peak Area Increase (PAI) of carbonyl compounds, measured via FTIR spectroscopy, and the Total Acid Number (TAN), indicating that the acidity in bulk-oxidized oils is primarily due to these oxidation products. This methodology provides insights into the degradation process of lubricants, emphasizing the role of acidic by-products in oil degradation.

2. **Increased Sludge Formation**: Yano et al. (2004) found that during oxidation, amine-type antioxidants in turbine oils degrade into insoluble, high-molecular-weight products that form sludge, which deposits on bearing surfaces, raises bearing temperatures, clogs filters, and accumulates in tanks. This sludge formation risk is heightened in low-aromatic hydrocracked base oils where degraded additives have poor solubility. The study highlights that oxidative degradation products—especially from PAN-type amines—can impair lubrication systems by fouling critical components and restricting oil flow. The article "Impact of Sludge on Hydraulic and Lubricating Systems" highlights the significant but often underestimated issues caused by sludge contamination in hydraulic and lubrication systems. Sludge, a by-product of oil degradation, accelerates wear by trapping abrasive contaminants and impairs system performance by clogging filters, reducing heat exchanger efficiency, and causing valve malfunctions. It also restricts lubrication flow, forcing systems to operate under boundary lubrication conditions, which increases wear on critical components like pumps, bearings, and gears. Additionally, sludge promotes system corrosion through the formation of acidic by-products and accelerates oil degradation, further compromising reliability. The article emphasizes the importance of proactive sludge removal and system maintenance to prevent these issues and maintain operational efficiency. Chokelarb et al.'s study introduces a Performance Rating Index (PRI) that consolidates metrics such as membrane patch colorimetry, viscosity, foaming stability, water separability, air release, oxidation stability, rust prevention, and copper strip corrosion to accurately predict the remaining useful life (RUL) of used lubricating oil.

1.5.3 Efficiency and Energy Consumption

The presence of foam and aeration in lubricants and coolants can adversely affect the efficiency of the electric drive unit, leading to increased energy consumption and reduced vehicle range. Understanding these impacts is critical for optimizing the performance and efficiency of electric vehicles.

1.5.3.1 Influence on Power Loss and Energy Efficiency

1. **Increased Power Loss due to Aeration**: Aeration increases fluid compressibility, leading to power losses in pumps and reduced volumetric efficiency. The increased compressibility caused by entrained air leads to cavitation and power loss in oil pumps, increasing energy consumption by up to 10%

in high-aeration scenarios. Quantitative analysis performed by Mastrone et al. (2023) revealed that aerated lubricants resulted in approximately a 5% increase in churning losses compared to non-aerated conditions. This increase is attributed to the combined effects of reduced mixture viscosity and increased specific volume due to air entrainment, which together affect the gearbox's efficiency.

2. **Hydraulic and Churning Losses**: The presence of aerated lubricants can impact the efficiency of the gearbox and driveline in complex ways. While entrained air bubbles can disrupt the lubricant's flow and increase its apparent viscosity, this does not always lead to higher energy losses. According to Mastrone et al. (2023), aeration alters oil distribution inside the gearbox, leading to increased splashing and changes in lubricant levels, particularly at higher rotational speeds. Their simulations showed that aeration reduces viscous losses, as air decreases the fluid's overall resistance to shear. However, this reduction in viscous losses is counterbalanced by an increase in inertial losses due to the air's effect on flow dynamics. For example, at a 10% aeration level, viscous losses were reduced by 10%, but inertial losses increased by an equivalent amount, particularly at lower speeds. This shift became more pronounced at higher speeds, with up to a 15% change in the balance between viscous and inertial losses. While aerated fluids can increase the system's resistance to motion by altering flow characteristics, the net effect on energy consumption depends on the operational conditions, such as speed and the level of aeration.

1.5.3.2 Effects on Overall EV Performance and Range

- **Reduced EV Range due to Increased Energy Consumption**: The study "Prediction of Electric Vehicle Transmission Efficiency Using a New Thermally Coupled Lubrication Model" (Shore et al., 2022) examines how transmission losses in electric vehicles (EVs) can significantly impact overall efficiency and range. It introduces a method to predict power losses in EV transmission systems by considering factors such as gear friction, bearing losses, and oil temperature. Electric motors in EVs are far more efficient than internal combustion engines (90% vs. ~40%), but this means that transmission losses, though small in absolute terms, represent a larger share of total power losses in EVs—about 15%–20%. By better predicting and reducing these losses through optimized gearbox design and lubricant selection, EV efficiency and battery range can be significantly improved. Using the introduced method, the study indicates that transmission losses can account for 15%–25% of the total power losses in an EV, a notably higher proportion compared to internal combustion engine vehicles. This increased energy consumption due to transmission inefficiencies directly reduces the vehicle's range. By optimizing gearbox design and lubricant properties, the study suggests that these losses can be minimized, thereby enhancing EV efficiency and extending range.

- **Impaired Thermal Management and Battery Efficiency**: While specific studies directly linking foam formation in cooling systems to elevated operating temperatures in power electronics are limited, existing research on coolant aeration provides valuable insights. For instance, the study "Numerical Modelling of Coolant Filling and De-aeration in a Battery Cooling System" examines the adverse effects of trapped air bubbles on cooling performance. The presence of air pockets can impede effective heat transfer, leading to increased operating temperatures in components such as power electronics. This research underscores the importance of effective de-aeration processes to maintain optimal thermal management. Additionally, the article "A Critical Review and Perspective on Thermal Management of Power Electronics Modules for Inverters and Converters" discusses various cooling technologies for power electronics, including liquid cooling methods. While it doesn't specifically address foam formation, the study highlights that any inefficiencies in the cooling system can lead to elevated junction temperatures, adversely affecting performance and reliability.

These impacts highlight the critical need for effective fluid management strategies to reduce foaming and aeration, ensuring the efficient and reliable operation of EDUs in electric vehicles.

1.6 TESTING AND CHARACTERIZATION OF FOAMING AND AERATION IN EV FLUIDS

The accurate assessment of foaming and aeration in electric vehicle (EV) lubricants and coolants requires specialized testing methodologies that replicate real-world conditions found in electric drive units (EDUs). Traditional testing methods, such as ASTM D892 (Standard Test Method for Foaming Characteristics of Lubricating Oils) and ASTM D3427 (Air Release Value of Petroleum Oils), have limitations that make them inadequate for characterizing foaming and air entrainment under the extreme conditions present in EDUs. This section outlines the experimental techniques used for foaming assessment, their limitations, and newer methodologies developed to address these challenges (Figure 1.4).

1.6.1 Experimental Techniques for Foaming Assessment

1.6.1.1 High-Speed Imaging and Analysis of Foam Formation

High-speed imaging is a valuable tool for visualizing and analyzing foam formation and air entrainment in EV fluids under dynamic conditions. This technique involves capturing rapid sequences of images at high frame rates, which allows researchers to observe the formation, growth, and collapse of foam bubbles in real time. High-speed imaging provides insights into bubble dynamics, foam stability, and the interaction between air bubbles and fluid additives.

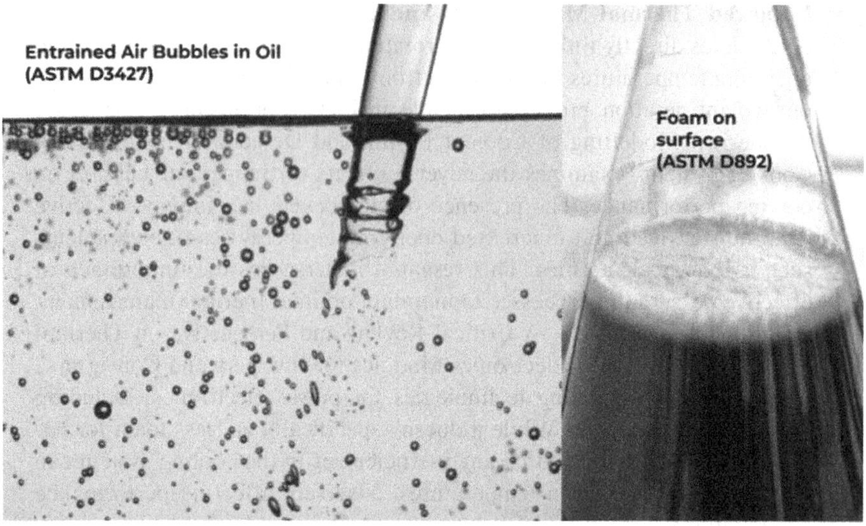

FIGURE 1.4 ASTM D892 & ASTM D3427. (Images from iStock.)

1.6.1.1.1 Applications in EV Fluids Testing

- In a study by S. Peerzada (personal communication, May 2023), high-speed imaging was used to capture the interaction of entrained air bubbles in lubricants under varying shear rates and temperatures (A Study of the Effects…). This visualization revealed how different anti-foaming agents affected the stability of foam structures in EV transmission fluids.
- High-speed imaging was also employed to quantify foam height, bubble size distribution, and the rate of bubble collapse, enabling a detailed analysis of foam stability under high-speed and high-temperature conditions similar to those found in EDUs.

1.6.1.1.2 Advantages

- Real-time observation of foam dynamics, providing a comprehensive view of bubble formation, coalescence, and collapse.
- The ability to simulate real-world operating conditions, such as varying shear rates and temperatures.

1.6.1.1.3 Limitations

- High-speed imaging is primarily qualitative and may require supplementary quantitative techniques to provide a complete characterization of foam properties.

1.6.1.2 Rheological Testing and Measurement of Foam Stability

Rheological testing measures the flow properties and stability of foamed lubricants and coolants under varying shear rates and temperatures. This technique provides insight into how foam formation affects the fluid's viscosity and elasticity, both of which influence lubrication and cooling performance.

1.6.1.2.1 Use of High Shear-Air Sparge Test

The High Shear-Air Sparge Test, as described by S. Peerzada (personal communication, May 2023), is a modified method designed to overcome limitations of traditional ASTM testing. This innovative approach evaluates fluid behavior under high-stress conditions, simulating the extreme environments experienced by electric vehicle (EV) fluids. By combining air sparging with high shearing, the test offers critical insights into foam formation, stability, and air entrainment.

1.6.1.2.2 Overview and Equipment

The test utilizes an Ultra-Turrax device, a high-performance mixing and homogenization tool that operates through a rotor-stator system:

- The **rotor**, spinning at high speeds (up to tens of thousands of revolutions per minute), generates intense shear forces.
- The **stator** works in tandem with the rotor to create rapid dispersion, emulsification, and homogenization of the fluid sample.

The rotor speed can be adjusted to suit specific sample types and desired outcomes, ranging from low-speed mixing to high-speed emulsification. This adaptability makes the test suitable for replicating a range of operating conditions.

1.6.1.2.3 Test Procedure

The High Shear-Air Sparge Test follows these steps:

1. **High-Speed Shearing**:
    - The sample is subjected to high-speed shearing using the Ultra-Turrax device for a specified duration.
2. **Air Sparging**:
    - Air is introduced into the fluid during shearing to simulate aeration effects.
3. **Volume and Entrained Air Measurement**:
    - Foam volume (FV) and entrained air (EA) are measured after mixing to assess foam formation and stability.

1.6.1.2.4 Results

During the test ($T=0$ to $T=300$ seconds), air sparging and shearing lead to foam formation, with foam volume and entrained air increasing over time. In the subsequent foam knockdown phase ($T=1$ to $T=60$ seconds), the foam collapses, and entrained air is released, mimicking real-world air release behavior (S. Peerzada, personal communication, May 2023).

1.6.1.2.5 Adjustable Parameters

The test allows customization to replicate specific conditions. The following parameters can be tailored:

- **Shear speed**: To simulate different mechanical stress levels.
- **Temperature**: To mimic operating conditions of electric drive units (EDUs).
- **Airflow rate**: To control the level of aeration.
- **Duration**: To simulate extended exposure to high-stress environments.

1.6.1.2.6 Applications and Relevance

This test is highly valuable for evaluating the performance and stability of EV fluids under high-stress conditions. By analyzing foam formation, stability, and air entrainment, researchers can:

- Optimize fluid formulations.
- Test and compare defoamer chemistries.
- Improve the reliability and performance of EV fluids in extreme operating conditions.

1.6.1.2.7 Advantages

- Provides quantitative data on foam volume, stability, and air entrainment under controlled shear and temperature conditions.
- Enables evaluation of various defoamer chemistries to identify the most effective formulations for foam reduction.

1.6.1.2.8 Limitations

While the High Shear-Air Sparge Test addresses several limitations of traditional ASTM methods, it remains a fluid-level test. It does not replicate the in-situ conditions of an actual EDU environment, such as complex geometries, varying pressures, or surface interactions with components.

1.6.1.2.9 Rheological Insights

The test integrates rheological evaluation by analyzing fluid behavior under dynamic shear conditions:

- **Viscosity**: Changes in foam formation and bubble behavior provide insights into how foam affects flow characteristics at varying shear rates.
- **Elasticity**: Bubble deformation and recovery highlight the elasticity of foam films, which can influence foam stability.
- **Foaming Impact**: By measuring foam stability over time, the test reveals how foam alters the fluid's overall rheological properties, such as viscosity and flow uniformity, crucial for lubrication and cooling systems.

1.6.1.3 Limitations of Traditional ASTM Tests

Traditional ASTM tests, such as ASTM D892 and ASTM D3427, are widely used for evaluating the foaming characteristics and air release properties of lubricating oils. ASTM D892 is a foaming characteristics test, which measures the volume of air bubbles as they rise and accumulate at the top of the sample, as well as the duration of their stability. ASTM D3427 is an air release test, which evaluates the speed at which air bubbles rise and escape from an oil sample (Savant Labs, 2024). However, these methods have significant limitations when applied to the unique operating conditions of EV EDUs:

1. **ASTM D892 (Foaming Characteristics of Lubricating Oils)**:
   - **Does Not Simulate Real-World Conditions**: ASTM D892 tests foam formation under controlled, low-shear conditions that do not reflect the high-speed, high-temperature, and high-pressure environments seen in EDUs. As a result, the test often fails to capture the true foaming behavior of the fluid.
   - **Additive Interference**: During ASTM D892 tests, the use of porous stones to introduce air can strip out some of the additives in the lubricant, leading to erroneous readings that do not accurately represent the fluid's performance in real applications (A Study of the Effects…) (Figure 1.5).
2. **ASTM D3427 (Air Release Value of Petroleum Oils)**:
   - **Limited to Low-Shear Applications**: This test measures the time required for entrained air to separate from the fluid under low-shear conditions. However, the extreme shear and turbulence present in EDUs

FIGURE 1.5 Tamson ASTM D892 foaming dual bath. (Reprinted with permission from PM Tamson Instruments, 2025.)

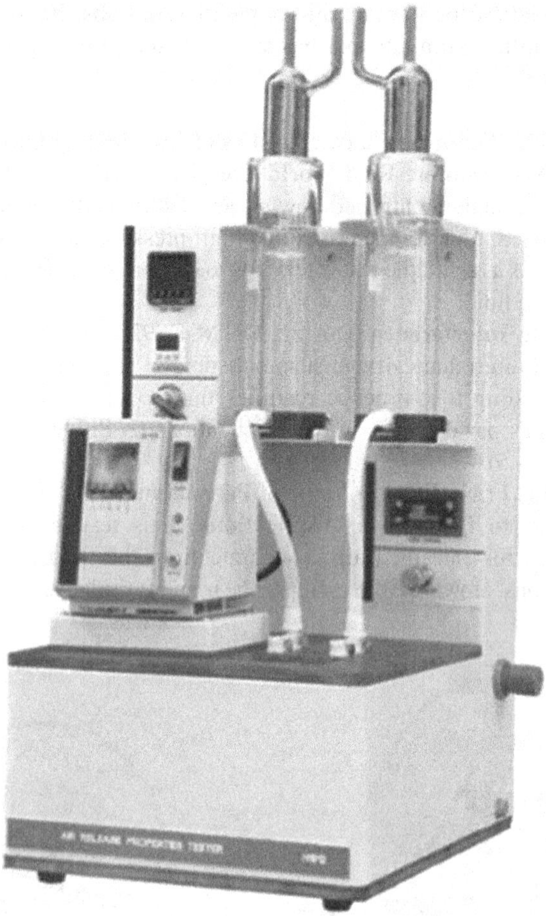

FIGURE 1.6 ASTM D3427 automated air release properties analyzer of hydrocarbon-based oils. (Reprinted with permission from Chongqing Gold Mechanical & Electrical Equipment Co., Ltd., 2025.)

can alter the fluid's air release behavior, making ASTM D3427 less relevant for evaluating the performance of EV fluids (A Study of the Effects…) (Figure 1.6).

3. **Porous Stone Artifacts**: The use of porous stones in these tests can introduce inconsistencies, as the stones may alter the fluid's composition by stripping out key additives or introducing air at a rate that does not reflect actual operating conditions. This issue can lead to an underestimation of the fluid's foaming tendency or air release performance.

1.6.1.4 New Testing Methods: High Shear-Air Sparge Test

The High Shear-Air Sparge Test was developed to address the limitations of traditional ASTM tests by replicating the high-shear conditions typical of conventional

EDUs. While it is capable of simulating high temperatures to assess foam formation and stability under extreme conditions, it can also be adjusted to reflect the lower fluid temperatures observed in many modern EDUs, as noted in recent testing practices. This adaptability ensures that the test is relevant for a wide range of operating environments, from conventional transaxles to lower-temperature EDUs. However, it is still a fluid-level test rather than an in-situ EDU test, meaning it does not account for interactions with surfaces, complex fluid dynamics, or component-specific effects that may influence foam formation and stability.

1.6.1.4.1 Advantages over ASTM Tests

- The High Shear-Air Sparge Test provides a more realistic assessment of foam behavior under high-shear conditions, making it more representative of the extreme environments found in EDUs.
- The ability to control shear speed, air injection rate, and temperature allows for a more tailored evaluation of fluid performance and the effectiveness of defoamer additives.

1.6.1.4.2 Remaining Limitations

- Despite its advantages, the High Shear-Air Sparge Test remains a fluid-level test and may not fully capture the complex in-situ conditions found in EDUs. For example, it does not consider the interactions between the fluid and the surfaces of gears, bearings, or other EDU components.

1.6.1.4.2.1 Conclusion While traditional ASTM testing methods like ASTM D892 and ASTM D3427 provide a baseline for evaluating foaming and air release properties, they do not adequately represent the extreme conditions present in EV EDUs. The High Shear-Air Sparge Test offers an improved approach by simulating high-shear and high-temperature environments, but it still falls short of replicating in-situ EDU conditions. Advanced in-situ testing methods that incorporate real EDU components and simulate actual operating conditions will be crucial for fully understanding the impact of foaming and aeration on EV fluid performance.

1.6.2 Air Entrainment Measurement Techniques

Accurate measurement of air entrainment and foaming in lubricants and coolants is crucial for understanding their effects on electric drive units (EDUs) and ensuring reliable performance under real-world conditions. Various techniques have been developed to quantify air content in fluids, each with unique working principles, advantages, and limitations. The following sections discuss the most widely used air entrainment measurement techniques, their operational mechanisms, and their pros and cons, as sourced from experimental data and industry standards.

1.6.2.1 Coriolis Measurement Method

The Coriolis measurement method is commonly used to measure the density and flow rate of fluids. This method operates by passing the fluid through a vibrating tube, and the resulting changes in the tube's oscillation frequency are used to calculate the fluid's density. When air is entrained in the fluid, it reduces the overall density of the mixture, and the Coriolis meter can detect these changes in real time.

1.6.2.1.1 How It Works

- The Coriolis meter measures the mass flow rate and density of a fluid simultaneously. By analyzing variations in density, it can determine the percentage of entrained air within the fluid.
- The sensor consists of a U-shaped or straight tube that vibrates at a specific frequency. When a fluid flows through the tube, the Coriolis effect induces a phase shift in the oscillation, proportional to the mass flow rate.

1.6.2.1.2 Pros

- Provides continuous, real-time measurement of fluid density and flow rate.
- Can be integrated into process loops for high-flow-rate applications, making it ideal for testing EDUs in real-time operating conditions.
- Capable of measuring entrained air over a wide range of flow rates and temperatures.

1.6.2.1.3 Cons

- The accuracy of the Coriolis method decreases at higher aeration rates, as the assumption of a homogeneous fluid breaks down with increasing air content.
- Requires careful calibration to differentiate between density changes due to temperature variations and those caused by aeration.

1.6.2.2 Impedance/Electrical Measurement Method

The impedance or electrical measurement method is based on the principle of detecting changes in the electrical properties of the fluid, such as conductivity or capacitance, as air becomes entrained. This method uses two electrodes placed inside a tube, with one acting as a signal generator and the other as a receiver.

1.6.2.2.1 How It Works

- As air is entrained in the fluid, it alters the electrical admittance (Y) or another intrinsic electrical property of the composite fluid between the electrodes. The method compares the admittance of the fluid to that of air and calculates the air content.
- The measured impedance change is proportional to the amount of air entrained within the fluid, and the signal can be used to continuously monitor aeration levels.

1.6.2.2.2 Pros

- Allows for continuous measurement with a fast data acquisition rate (e.g., 100 ms), making it suitable for detecting transient aeration events.

- Not significantly affected by temperature variations, making it more stable across a wide range of operating conditions.

1.6.2.2.3 Cons

- Requires calibration for each fluid type due to variations in fluid conductivity and other electrical properties.
- High levels of conductive contamination, such as water or metal particles, can affect the measurement accuracy.

1.6.2.3 X-Ray Absorption Method

The X-ray absorption method is a non-destructive technique used to measure the density of aerated fluids. This method relies on the fact that air does not absorb X-rays, while oil or coolant fluids do. By measuring the attenuation of X-rays through the fluid, the method can determine the density and, consequently, the air content.

1.6.2.3.1 How It Works

- A sample of the fluid is placed between an X-ray source and a detector. As X-rays pass through any fluid, they are absorbed based on the density of the material. When the X-rays pass through the fluid and the entrained air, they are predominantly absorbed by the fluid due to its high density compared to that of the entrained air.
- The intensity of the transmitted X-rays is measured, and the change in intensity is used to calculate the air content by comparing it to the known densities of the fluid and air. To ensure accurate measurements, the system is calibrated beforehand by recording X-ray radiation yields for two reference conditions: pure air and unaerated fluid.

1.6.2.3.2 Pros

- Offers high accuracy in measuring the density of the fluid, including distinguishing between dissolved air and entrained air.
- Data acquisition is quick, typically taking only a few seconds per measurement.

1.6.2.3.3 Cons

- Requires a sealed radioisotope source, making it less practical for routine measurements due to regulatory restrictions.
- High equipment costs and safety concerns associated with handling radioactive materials limit its application in some testing environments.

1.6.2.4 Compression or SMAC Method

The Système de Mesure d'Aération par Compressibilitié (SMAC) method measures the volumetric gas ratio of an aerated fluid by compressing a sample under controlled pressure and temperature conditions. The change in volume as the fluid is compressed is used to calculate the amount of entrained air.

1.6.2.4.1 How It Works

- A fluid sample is placed in a chamber, and the pressure is increased while the temperature is kept constant. The volume change in the fluid is monitored, and the difference between the initial and final volumes is used to determine the amount of air.
- The method is effective at quantifying air content up to a maximum aeration level of 15%. This method is limited to low aeration percentages because, at higher aeration levels, the gas content in the fluid becomes significant, and the compressibility of the gas begins to dominate the system's behavior. The relationship between pressure and volume in the gas phase becomes nonlinear, deviating from the assumptions used in the SMAC Method. This can lead to inaccuracies in calculating aeration percentages. Additionally, higher aeration levels often lead to bubble coalescence (merging of bubbles), resulting in uneven distribution of gas in the fluid. This creates heterogeneous conditions that are not well-suited to the assumptions of the SMAC Method, which assumes uniform distribution.

1.6.2.4.2 Pros

- Offers high accuracy for low-to-moderate aeration levels.
- Capable of distinguishing between different forms of aeration, such as dissolved air and entrained air.

1.6.2.4.3 Cons

- Limited measurement range, with accuracy decreasing at higher aeration levels.
- Cannot capture transients or dynamic aeration events since it requires sample collection and static testing.

1.6.2.5 Beaker Method

The beaker method is a simple and inexpensive technique used to measure air entrainment in fluids by collecting a sample and allowing the air to separate from the fluid over time. This method is particularly useful for preliminary assessments but has several limitations.

1.6.2.5.1 How It Works

- A fluid sample is collected in a graduated cylinder, and the initial volume (V_m) is recorded. After allowing sufficient time for entrained air to escape, the volume of the fluid is measured again (V_o).
- The percentage of air entrainment (A_{wo}) is calculated using the formula:

$$A_{wo} = \frac{V_m - V_o}{V_m} \times 100\% \tag{1.20}$$

1.6.2.5.2 Pros

- Low cost and easy to implement without specialized equipment.
- Provides a basic assessment of air content in fluids, useful for initial testing.

1.6.2.5.3 Cons

- Inaccurate and highly sensitive to temperature variations, leading to poor repeatability.
- Cannot measure dynamic aeration levels or capture transient events, making it unsuitable for real-time monitoring.

1.6.2.5.3.1 Summary Each air entrainment measurement technique has unique strengths and weaknesses, making them suitable for different applications depending on the test environment and desired data resolution. While traditional methods like the beaker test are inexpensive and straightforward, they lack accuracy and repeatability. Advanced methods like the Coriolis and impedance techniques provide real-time measurement capabilities but require careful calibration and may suffer from decreased accuracy at high aeration levels. Understanding these limitations is essential for selecting the appropriate method to evaluate fluid performance in electric drive units under various operating conditions.

1.6.3 Bubble Size Distribution and Gas Content Analysis

Understanding bubble size distribution and gas content analysis is crucial for evaluating the impact of aeration and foaming on electric drive units (EDUs). These parameters influence lubrication film thickness, thermal management, and the overall efficiency of the EDU system. Analyzing dissolved and entrained gas content provides insights into different aspects of fluid behavior and performance. While dissolved gas may not be a significant concern in a stable fluid state, its potential to come out of solution during operation is what makes it relevant to engineers. This section explores how bubble size distribution and gas content analysis are performed, with a focus on the implications for EDU performance.

1.6.3.1 Analysis of Bubble Size Distribution in EDUs

Bubble size distribution within a lubricant or coolant plays a significant role in determining how the fluid behaves under different operating conditions. In EDUs, both small and large bubbles can have detrimental effects on lubrication and heat transfer performance:

- **Small Bubbles (<100 microns)**: These bubbles tend to remain suspended in the fluid for longer periods due to their low buoyancy. They are difficult to separate and can accumulate in areas of high turbulence, such as around gears and bearings. The presence of small bubbles can alter the fluid's effective viscosity and density, disrupting the lubrication film and potentially leading to increased wear.

- **Large Bubbles (>500 microns)**: Large bubbles, while more buoyant and easier to separate from the fluid, can cause rapid changes in fluid volume and pressure. Their collapse can lead to intermittent lubrication breakdowns and cavitation, damaging surfaces and reducing component lifespan.

1.6.3.1.1 Measurement Techniques

- **Optical Imaging**: High-speed cameras can capture real-time images of bubble formation and coalescence. Using particle tracking analysis, the size and distribution of bubbles can be quantified under varying shear and temperature conditions.
- **Ultrasonic Sensors**: Ultrasonic waves interact differently with bubbles of varying sizes, providing data on bubble concentration and distribution in opaque fluids.

1.6.3.1.2 Relevance in EDUs

The presence of a bimodal bubble size distribution, where both small and large bubbles are present, is common in high-shear environments like EDU gearboxes. This distribution pattern affects how the fluid distributes load, dissipates heat, and maintains lubrication film stability under high-stress conditions.

1.6.3.2 Gas Content Analysis and Its Impact on EDU Performance

While high levels of dissolved gas in a sump or static fluid condition may not be a significant concern, they become critically important when they come out of solution due to pressure or temperature changes. Once dissolved gases are released, they can form bubbles that increase fluid aeration, reducing lubrication and thermal transfer effectiveness.

1.6.3.2.1 Measurement Techniques for Dissolved Gas Content

1. **Optical Sensors for Dissolved Gas**: Optical sensors measure the concentration of dissolved gases such as oxygen or nitrogen by detecting changes in light absorption or scattering as gases dissolve in the fluid. These sensors are effective for continuous monitoring and can indicate when dissolved gases are nearing saturation levels that could lead to bubble formation.
    - **Relevance to EDU Performance**: Dissolved gases are not inherently harmful while in solution. However, when they come out of solution—typically due to sudden pressure drops or temperature spikes—they form bubbles that disrupt fluid behavior. For example, a lubricant with high dissolved gas content could suddenly release these gases when entering a low-pressure region in a gearbox, leading to cavitation and lubrication film instability. Therefore, monitoring dissolved gas content helps predict and prevent potential aeration issues.
2. **Gas Chromatography**: Gas chromatography is a laboratory technique used to separate and quantify dissolved gases in a fluid. This method is highly accurate and can measure the concentrations of individual gases, such as oxygen, nitrogen, and carbon dioxide. It is especially useful for analyzing how dissolved gas concentrations change under different operating conditions and how these changes correlate with fluid performance.

- **Relevance to EDU Performance**: Understanding the concentration of specific dissolved gases can help in identifying the root causes of oxidation and chemical degradation. For example, high levels of dissolved oxygen can promote oxidative reactions that produce acidic by-products, leading to increased corrosion and reduced lubricant lifespan.

1.6.3.2.2 Impact of Dissolved Gas on EDU Performance

While most concerns related to fluid performance focus on entrained air and its direct impact on compressibility and lubrication, dissolved gases should not be overlooked. High dissolved gas content can lead to:

- **Formation of Microbubbles**: When dissolved gases come out of solution, they often form microbubbles that can significantly disrupt lubrication film stability. These bubbles may coalesce into larger ones, increasing aeration and foaming.
- **Increased Risk of Cavitation**: Dissolved gases released from solution in low-pressure regions can cause cavitation, leading to pitting and erosion of surfaces, especially in high-speed applications such as electric drive gearboxes.
- **Oxidation and Chemical Degradation**: Dissolved oxygen and other reactive gases can accelerate oxidation, leading to the formation of sludge, varnish, and acidic by-products that degrade the fluid's thermal and lubricating properties over time.

1.6.3.2.3 Practical Implications

While monitoring dissolved gas content may not always be necessary under stable sump conditions, it becomes critically important in dynamic environments where pressure and temperature fluctuations are common. Effective dissolved gas management ensures that the fluid maintains its designed properties and does not introduce unwanted bubbles that compromise lubrication and thermal management. For instance, knowing the threshold at which a fluid releases its dissolved gases can help in designing systems that avoid crossing these critical pressure or temperature points, thereby preventing sudden aeration and ensuring consistent EDU performance.

1.7 SIMULATION AND MODELING APPROACHES

Accurate prediction and analysis of foam formation and aeration in electric drive units (EDUs) are critical for understanding fluid behavior and optimizing system performance. Given the complexity of fluid dynamics in these systems, traditional experimental methods may not always provide a complete picture of fluid behavior. Simulation and modeling approaches, such as computational fluid dynamics (CFD) and multiphase flow simulations, have become valuable tools for predicting and analyzing foam and aeration phenomena. This section discusses how these advanced simulation techniques can be applied to better understand foam and aeration behaviors within EDUs.

1.7.1 Computational Fluid Dynamics (CFD) Models for Foam Prediction

Computational Fluid Dynamics (CFD) is a powerful simulation tool used to predict fluid behavior, including foam formation and stability, under various operating conditions. CFD models solve the governing equations of fluid flow, such as the Navier-Stokes equations, to provide detailed insights into the velocity, pressure, and temperature fields within a system. In the context of foaming in EDUs, CFD models can be adapted to include specialized sub-models that account for foam dynamics, including bubble coalescence, breakup, and drainage.

1.7.1.1 Key Aspects of CFD Foam Prediction

1. **Modeling Bubble Dynamics**: CFD simulations can incorporate Volume of Fluid (VOF) models or Eulerian-Eulerian approaches to track the interface between the gas and liquid phases. These models are used to simulate the behavior of individual bubbles or clusters of bubbles, including their rise velocity, coalescence, and breakup in high-shear regions, such as those found in EDU gearboxes.
2. **Foam Formation and Stability Models**: CFD can include sub-models to predict foam formation based on surface tension effects, interfacial forces, and rheological properties of the fluid. For example, the Homogeneous Multiphase (HMP) model or Population Balance Model (PBM) can be used to account for the formation, growth, and collapse of foam bubbles.
3. **Effect of Geometry and Flow Conditions**: CFD simulations allow for the analysis of how specific geometric features, such as sharp corners or narrow channels, affect foam formation and stability. By adjusting the geometry and flow conditions in the simulation, engineers can identify design modifications that minimize foam generation.

1.7.1.2 Advantages

- CFD provides detailed spatial and temporal resolution of fluid flow, making it possible to study the formation and behavior of foam at different locations within the EDU.
- The ability to simulate different operating conditions, such as varying flow rates, temperatures, and pressures, helps predict foam behavior under realistic scenarios.

1.7.1.3 Limitations

- High computational cost and complexity, particularly when modeling multi-phase interactions with fine-scale bubble dynamics.
- Requires accurate input data for fluid properties, surface tension, and empirical correlations for foam stability, which may not always be readily available.

1.7.2 Multi-Phase Flow Simulations for Aeration Analysis

Multi-phase flow simulations are used to analyze the behavior of fluids containing more than one phase, such as liquid and gas. In the context of aeration analysis in

EDUs, multi-phase simulations can model the interactions between the liquid lubricant or coolant and the entrained gas bubbles. These simulations provide insights into how air is introduced, transported, and distributed within the fluid, and how it affects overall system performance.

1.7.2.1 Key Approaches in Multi-Phase Flow Simulations

1. **Eulerian-Eulerian and Eulerian-Lagrangian Models**:
   - **Eulerian-Eulerian Models**: In these models, both the liquid and gas phases are treated as interpenetrating continua. This approach solves separate sets of conservation equations for each phase, including mass, momentum, and energy, allowing for the prediction of volume fraction distributions, velocity fields, and pressure changes due to aeration.
   - **Eulerian-Lagrangian Models**: These models track the motion of individual gas bubbles (Lagrangian phase) as they move through the continuous liquid phase (Eulerian phase). This approach provides detailed information on bubble trajectories, interactions with surfaces, and coalescence or breakup events.
2. **Multi-Component Flow Models**: Multi-component flow models can simulate fluids with different compositions, such as oil and gas mixtures. By including sub-models for mass transfer between phases, these simulations can predict when dissolved gases will come out of solution and form bubbles, as well as how bubbles will distribute within the flow.
3. **Cavitation and Bubble Dynamics Models**: Multiphase flow simulations can include cavitation models that predict the formation and collapse of vapor-filled cavities (bubbles) in regions of low pressure. These models are particularly relevant in high-speed applications, where sudden pressure drops can cause dissolved gases to come out of solution, leading to cavitation and potential damage to EDU components.

1.7.2.2 Advantages

- Multi-phase flow simulations provide a comprehensive view of how entrained air and dissolved gases behave under varying pressure and temperature conditions.
- The ability to model complex interactions between phases allows for a more accurate prediction of aeration effects on lubrication and thermal management.

1.7.2.3 Limitations

- Multi-phase flow simulations are highly sensitive to the choice of models and numerical schemes, making them complex to set up and validate.
- Like CFD, these simulations are computationally intensive and require high-quality input data to produce reliable results.

1.7.2.3.1 Relevance to EDU Applications

Simulation and modeling approaches, such as CFD and multiphase flow simulations, are essential for predicting and mitigating the negative effects of foam formation and aeration in EDUs. These techniques enable engineers to evaluate how different design parameters and operating conditions influence fluid behavior, allowing for the optimization of lubricant and coolant formulations as well as system geometries. While these models provide valuable insights, they must be validated with experimental data to ensure accuracy and reliability. Combining simulations with empirical testing allows for a comprehensive understanding of foam and aeration dynamics, ultimately leading to more robust and efficient EDU designs.

1.8 CASE STUDIES

Foaming and aeration issues in electric drive units (EDUs) have been a persistent challenge in the automotive electrification industry. These problems, often discovered during testing or real-world operation, can severely affect lubrication efficiency, thermal management, and overall performance. This section discusses real-world case studies of foaming and aeration issues encountered by leading EV manufacturers and presents insights from recent research studies that highlight the importance of addressing these issues.

1.8.1 Real-World Examples of Foaming and Aeration Issues in EDUs

The following examples are drawn from real-world experiences and past professional knowledge accumulated during the author's work in the field of precision fluid systems engineering. They highlight critical challenges, problem-solving approaches, and lessons learned in various projects to provide readers with practical insights and a deeper understanding of the complexities encountered in engineering, technology development, and problem-solving in dynamic, real-world scenarios. To maintain confidentiality and respect the privacy of all involved parties, specific names, brands, and identifying details have been anonymized or generalized.

1. **Leader in EVs Drive Unit Noise and Durability Issues**: Early versions of the car made by a Leader in EVs experienced significant foaming and aeration issues within their drive units. During aggressive driving scenarios or when operating at high speeds, the lubricating oil in the gearbox would aerate, leading to the formation of foam. This caused lubrication film breakdown, leading to increased friction and premature wear of the gears and bearings. The resulting noise and vibration issues were exacerbated by the loss of lubrication efficiency, leading to multiple drive unit replacements under warranty.

   The leader in EVs addressed this issue through several iterations of hardware changes, including adjustments to fluid formulations, redesigning the lubrication paths to minimize turbulence, and adding anti-foaming additives to the gearbox oil. These changes significantly reduced aeration levels and improved the durability of the drive unit. However, the issue was widely publicized and highlighted the challenges faced in designing robust lubrication systems for high-performance EVs.

2. **Leader in Truck EVs Thermal Management Challenges**: Leader in Truck EV models were reported to have experienced thermal management issues in early production versions, partly due to aeration within the cooling and lubrication systems. The intense off-road driving and towing capabilities of these vehicles led to the formation of foam in the transmission oil during high-load conditions. Foam formation impaired the oil's ability to transfer heat, resulting in increased operating temperatures in the drive unit. The engineers addressed these problems by optimizing the gearbox design to reduce fluid agitation and implementing improved defoaming agents in the lubricant to stabilize the oil under extreme conditions.
3. **Chinese EV Gearbox Overheating Issue**: During the initial rollout of a leading Chinese EV manufacturer, drivers reported instances of gearbox overheating when operating the vehicle at high speeds for extended periods. Investigations revealed that the overheating was due to foaming in the gearbox oil, which reduced the oil's thermal conductivity and compromised heat dissipation. This issue was traced back to aeration caused by high levels of turbulence within the oil flow path. The Chinese EV manufacturer implemented a redesigned lubrication system with anti-foaming additives and added air-bleed mechanisms to alleviate the issue. This case illustrates the importance of proper aeration management in maintaining thermal stability and preventing overheating in EV drive units.
4. **German EV High-Speed Testing Failures**: A German EV manufacturer experienced lubrication-related failures during high-speed testing, attributed to air entrainment in the transmission oil. The entrained air increased fluid compressibility, leading to pressure drops and cavitation at the pump inlet. These failures were more frequent in cold climates, where the oil's viscosity was higher, making it more prone to trapping air. The German EV manufacturer addressed these issues by optimizing the pump design and using a low-viscosity synthetic oil with enhanced anti-foaming properties.

These case studies underscore the complexity of managing foaming and aeration in EV EDUs, especially under varying operating conditions. Effective solutions often require a combination of mechanical design changes and chemical modifications to the fluids.

1.8.2 Test Results and Insights from Recent Research

1. **Studies on Anti-Foaming Additives and Their Effectiveness**: Recent research has focused on evaluating the effectiveness of various anti-foaming additives in reducing foam formation and stabilizing lubrication films in high-performance EV applications. One such study conducted by Munzing and published in SAE International analyzed the performance of fluorosilicone-based defoamers compared to traditional silicone-based additives. The findings indicated that fluorosilicone-based defoamers provided a 30% reduction in foam height and a 20% improvement in film stability under high-shear conditions.

2. **Impact of Oil Flow Path Design on Aeration Levels**: Baran's thesis examines factors influencing oil aeration in internal combustion engines, including the design of the oil circulation system. It highlights that the routing of returning oil directly into the sump below the oil level can minimize contact with air in the crankcase, thereby reducing aeration. While the study doesn't provide a detailed analysis of oil flow path design, it underscores the importance of proper oil return pathways in controlling aeration levels.
3. **Development of Multi-Phase CFD Models for Foam Prediction**: Sattar et al.'s (2013) paper introduces a computational fluid dynamics (CFD) approach that incorporates population balance modeling to simulate the behavior of aerated liquids, focusing on creaming and foam formation. By treating foam as a distinct phase composed of both air and liquid, the model effectively captures the dynamics of bubble coalescence and breakup. The simulation results align well with experimental data, validating the model's accuracy. This methodology offers valuable insights into the mechanisms of foam formation and stability, which are crucial for optimizing industrial processes where foam presence affects efficiency and product quality.
4. **Experimental Analysis of Aeration Effects on Bearing Wear**: Again, the combined findings from Goodwin et al. (2007) and Chun highlight the detrimental impact of aeration on bearing performance. Goodwin et al. focus on hydrodynamic journal bearings and demonstrate that air bubbles in the lubricating oil reduce the load-carrying capacity due to increased oil compressibility and decreased viscosity, particularly at higher operating speeds. This research emphasizes the need to control aeration levels to maintain an effective hydrodynamic film and ensure optimal load support. Chun's study explores turbocharger journal bearings and similarly finds that aerated lubricants reduce load-carrying capacity, while also altering the bearing's dynamic behavior. These changes are attributed to the disruption of the oil film, which compromises the bearing's ability to handle dynamic loads efficiently.

These research studies provide valuable insights into the mechanisms of foam formation and aeration and offer evidence-based strategies for mitigating their effects in EDUs. By combining simulation models, empirical testing, and advanced additive formulations, the industry can continue to improve the reliability and efficiency of EV powertrains.

1.9 MITIGATION STRATEGIES FOR FOAMING AND AERATION

The development and application of mitigation strategies for foaming and aeration in electric drive units (EDUs) are essential for designing robust and efficient systems. Engineers designing new electric vehicle (EV) architectures must integrate these strategies into the overall system design to ensure optimal lubrication and thermal management. This section provides a detailed technical manual on design, additive, and operational considerations tailored to R&D engineering teams tasked with defining and optimizing EV architectures.

1.9.1 Design and Material Selection

Design and material selection are foundational to mitigating foaming and aeration in EDUs. Engineers must incorporate design elements that minimize mechanical agitation and choose materials and fluids that inherently resist foam formation.

1.9.1.1 Selection of Low-Foaming Lubricants and Coolants

The selection of appropriate lubricants and coolants is the first step in addressing foam formation and aeration. Low-foaming fluids are specifically engineered to minimize the incorporation of air and prevent the stabilization of foam within the fluid.

- **Base Fluid Chemistry**: Low-foaming synthetic fluids, such as polyalphaolefin (PAO) or ester-based lubricants, should be prioritized for their superior thermal stability, low surface tension, and shear resistance. Their properties enable them to maintain stable performance under high shear rates and varying temperature conditions commonly encountered in EDUs.
- **Thermal and Rheological Properties**: The thermal properties of the lubricant, including its specific heat capacity and thermal conductivity, should be optimized to manage heat dissipation without causing localized boiling or bubble formation. Rheologically, the fluid must maintain consistent viscosity over a wide temperature range to prevent the trapping of air in high-shear zones.
- **Anti-Foaming Characteristics**: Fluids with lower surface tension are less likely to form persistent foam. Selecting lubricants with built-in anti-foaming properties can preempt the need for extensive chemical additives, reducing complexity and potential compatibility issues.

1.9.1.2 Design Modifications to Minimize Mechanical Agitation

Design modifications to reduce turbulence, shear forces, and splashing are crucial in preventing the formation of foam and minimizing air incorporation in the fluid.

- **Flow Path Optimization**: Fluid flow paths should be engineered with smooth contours, avoiding sharp corners and sudden transitions that can induce turbulence and increase the risk of air entrapment. Gradual transitions between channels reduce the formation of vortices, which are primary sites for air incorporation.
- **Minimization of Splash and Aeration Points**: The configuration of gears, pumps, and bearings within the EDU should minimize splash zones. Ensure that moving parts are adequately submerged to prevent splashing and the subsequent entrapment of air. Additionally, optimize the location and orientation of fluid inlets and outlets to prevent air suction.
- **Implementation of Air-Bleed and Degassing Systems**: Incorporating air-bleed valves and degassing systems into the fluid circuit helps actively remove entrained air. These systems should be strategically placed at high

points within the fluid circuit, where air naturally accumulates, to ensure the continuous removal of air from the lubricant or coolant.

- **Material Compatibility and Surface Coatings**: Selecting materials with low surface energy or applying surface coatings that reduce bubble adhesion can further help in minimizing foam stability. Components should be designed with surface finishes that discourage bubble adherence and promote rapid bubble release.

1.9.2 Additive Solutions

Incorporating chemical additives into the fluid is a proactive approach to controlling foam formation and aeration. Selecting the right additives and understanding their interactions with the base fluid is critical to achieving consistent performance.

1.9.2.1 Use of Anti-Foaming Agents and Surfactant Management

Anti-foaming agents are specially designed to alter the surface properties of the fluid, destabilizing foam bubbles and promoting their rapid collapse. For R&D engineers, it is essential to evaluate the effectiveness of these agents under the specific operating conditions of the EDU.

- **Anti-Foaming Additives**: Silicone-based and fluorosilicone-based additives are commonly used to reduce foam stability. These additives act by spreading over the foam's surface, reducing surface tension, and causing bubble walls to rupture. Engineers should test these additives for compatibility with the lubricant's base chemistry to ensure they do not negatively impact other performance properties.
- **Surfactant Control**: The presence of surfactants, while beneficial for detergency and dispersion, can contribute to foam stability. Engineers should carefully control the concentration of surfactants to avoid excessive foaming. Surfactant management is particularly important when designing multi-functional fluids that serve both as lubricants and coolants.

1.9.2.2 Chemical Additives to Reduce Aeration Effects

In addition to anti-foaming agents, other chemical additives can be used to reduce the effects of aeration and stabilize the gas–liquid interface.

- **Viscosity Modifiers**: These additives ensure the fluid maintains its rheological properties even when small amounts of air are present. By stabilizing the fluid's viscosity, air bubbles are less likely to remain suspended within the flow.
- **Surface Tension Modifiers**: Reducing the surface tension of the fluid helps prevent the formation of stable foam structures and reduces the size of entrained bubbles. Surface tension modifiers can be particularly useful in high-shear regions of the EDU, where smaller bubbles are more likely to form.
- **Oxidation Inhibitors**: High levels of aeration can lead to oxidation and degradation of the fluid. Adding oxidation inhibitors helps maintain fluid stability, reducing the likelihood of sludge or varnish formation and ensuring consistent performance over the lifespan of the fluid.

1.9.3 Operational Strategies

Operational strategies encompass managing the temperature, pressure, and fluid dynamics within the EDU to reduce the likelihood of foaming and aeration. These strategies should be incorporated into the design phase to ensure fluid stability under all operating conditions.

1.9.3.1 Temperature and Pressure Management within EDUs

Fluctuations in temperature and pressure can cause dissolved gases to come out of solution, leading to aeration and cavitation. Effective management of these variables is crucial to maintaining fluid stability.

- **Temperature Gradients**: Implementing temperature control strategies, such as using heat exchangers and maintaining consistent thermal profiles, prevents localized hot spots that could lead to rapid gas release and bubble formation. For example, the integration of thermal management systems with phase-change materials can reduce temperature spikes, maintaining fluid stability under peak loads.
- **Pressure Stabilization**: Pressure control mechanisms, such as using pressure regulators and pressure relief valves, can prevent sudden changes that cause dissolved gases to escape. Variable-speed pumps can also be employed to maintain constant flow rates and minimize pressure variations within the fluid circuit.

1.9.3.2 Fluid Circuit Design Considerations

Fluid circuit design plays a critical role in managing foaming and aeration, and should be optimized during the initial system architecture phase. The residence time of fluid in areas with large fluid/air interfaces—such as expansion chambers, reservoirs, or turbulent regions at free surfaces—directly impacts gas entrainment and foam generation. Longer residence times at these interfaces provide more opportunity for air to be released from solution, increasing the likelihood of foam formation and entrapment. Balancing flow rates and designing geometries to minimize stagnant zones can reduce the time fluids spend at these interfaces, thereby mitigating air release and subsequent foaming.

- **Minimizing Nucleation Sites**: Nucleation sites, where bubbles form, should be minimized through the use of high-purity materials and surface coatings that discourage bubble formation. Ensuring a clean, contamination-free system reduces the number of potential nucleation points for bubble formation.
- **Incorporation of Pressure and Temperature Gradients**: Designing the fluid circuit to incorporate gradual pressure and temperature changes allows gases to remain dissolved and prevents the rapid formation of foam. Fluid systems should be modeled and simulated using Computational Fluid Dynamics (CFD) to predict high-risk areas for foaming and aeration and implement corrective design strategies.

By applying these design, additive, and operational strategies during the development phase, R&D engineers can effectively mitigate foaming and aeration in new EV architectures, ensuring optimal performance, reliability, and durability of electric drive units.

1.10 FUTURE TRENDS AND RESEARCH DIRECTIONS

The rapid evolution of electric drive units (EDUs) and the increased demand for efficient electric vehicles (EVs) have spurred research into advanced lubricant and coolant technologies, innovative testing methods, and sustainable fluid solutions. As the field of EV design continues to progress, new trends and research directions are emerging, focusing on optimizing fluid dynamics, enhancing system performance, and reducing environmental impact. This section delves into these future trends, highlighting advancements in fluid formulations, diagnostic tools, and sustainable practices that are likely to shape the industry over the next few years.

1.10.1 Advancements in Lubricant and Coolant Formulations

Future advancements in lubricant and coolant formulations are driven by the need to enhance thermal management, reduce foaming and aeration, and increase fluid longevity under the demanding conditions of EDUs. Emerging technologies in fluid additives and the application of nanotechnology hold great promise in meeting these objectives.

1.10.1.1 Emerging Technologies in Fluid Additives

The development of advanced fluid additives is focused on enhancing the anti-foaming, anti-wear, and thermal management capabilities of lubricants and coolants. Novel surfactants, defoamers, and viscosity modifiers are being engineered to maintain stability and performance under extreme shear and temperature conditions.

- **Hybrid Polymer-Based Additives**: Innovations in hybrid polymer chemistry are resulting in the creation of multifunctional additives that offer improved performance at lower concentrations. These additives enhance anti-foaming properties while simultaneously providing anti-wear and friction reduction benefits, making them suitable for high-performance EDUs. Ren et al.'s (2023) study discusses various antifoaming agents, including silicone-based, non-silicone-based, and compound defoamers, detailing their synthesis, mechanisms of action, advantages, and limitations. The authors emphasize the importance of selecting appropriate antifoaming agents to maintain the performance and longevity of lubricating oils. This research extends to the development of hybrid polymer-based additives for fluids, which combine the benefits of different materials to enhance antifoaming efficiency. By integrating polymers with nanoparticles or other functional groups, these hybrid additives can provide improved foam suppression, thermal stability, and compatibility with various fluid systems.
- **Self-Adaptive Additives**: These cutting-edge additives can modify their molecular structure in response to changing operating conditions. This

adaptability allows the fluid to maintain consistent viscosity and thermal properties, reducing the likelihood of foam formation under varying shear rates and temperatures. Cadix et al. (2021) discusses how associative microgels, with their self-adaptive shear-induced aggregation, effectively reduce fluid loss by forming aggregates that seal variable pore sizes. This property can extend to foaming control, as the microgels' ability to form adaptive networks could potentially stabilize or break foam structures by interacting with gas–liquid interfaces, reducing excessive foaming in fluid systems. Their tunable behavior makes them promising as multipurpose fluid additives in industrial applications.

- **Thermally Responsive Additives**: Thermally responsive additives change their behavior based on fluid temperature, activating or deactivating certain properties as needed. These additives help stabilize the lubricant or coolant at different thermal states, optimizing fluid performance under both steady-state and transient conditions. The article *"Review of the Perspectives and Study of Thermo-Responsive Polymer Gels and Applications in Oil-Based Drilling Fluids"* (Wang et al., 2023) explores the use of thermo-responsive polymer gels that undergo phase transitions in response to temperature changes. These gels enhance the adaptability of oil-based drilling fluids by improving rheological properties, controlling fluid loss, and maintaining stability under varying downhole conditions. Thermo-responsive additives can help stabilize lubricants or coolants by adjusting their structure and behavior based on temperature. At higher temperatures, these polymers can change their solubility or aggregation state, preventing phase separation or degradation in lubricants and coolants. Conversely, at lower temperatures, they can enhance viscosity or form networks to maintain stability and performance. This ability to dynamically adapt ensures consistent functionality across a wide range of operational conditions, making thermo-responsive additives valuable for maintaining the efficiency and reliability of mechanical systems.

1.10.1.2 Nanotechnology and Its Potential to Reduce Foaming/Aeration

Nanotechnology is increasingly being integrated into lubricant and coolant formulations to address foaming and aeration issues at a molecular level. The incorporation of nanoparticles can significantly enhance the fluid's properties, allowing for better control over gas solubility, surface tension, and thermal conductivity.

- **Nanoparticle-Enhanced Fluids**: Adding nanoparticles such as silica, graphene oxide, or carbon nanotubes to lubricants and coolants can dramatically alter the fluid's behavior under high shear and high-pressure conditions. Nanoparticles reduce foam stability by destabilizing bubble walls, thereby promoting faster bubble coalescence and collapse. The article "Enhancing Lubricant Properties by Nanoparticle Additives" (Shahnazar et al., 2016) examines the impact of incorporating nanoparticles into lubricants to improve their tribological performance. The study highlights that adding nanoparticles such as TiO_2 and CuO to base oils can significantly reduce friction and wear. For instance, the inclusion of TiO_2 nanoparticles in rapeseed oil resulted in a 15.2% reduction in the coefficient of friction. Similarly, dispersing modified nanoparticles into base oils achieved a friction coefficient reduction of approximately 16.24% at an optimal concentration of 0.1 wt.%. These findings demonstrate that nanoparticle additives can substantially enhance the lubricating properties of oils, leading to improved efficiency and longevity of mechanical systems

- **Functionalized Nanoparticles**: Functionalizing nanoparticles with chemical groups that interact specifically with the base fluid molecules enhances dispersion and stability. These functionalized particles can create a dynamic barrier that prevents the formation of stable foam structures, making them ideal for high-performance EV systems. The chapter, *"Influence of Surfactant Adsorption on Surface-Functionalized Silica Nanoparticles for Gas Foam Stability"* (Hurtado et al., 2021), explores how the interaction between surface-functionalized silica nanoparticles and surfactants impacts foam stability. The study highlights that the adsorption of surfactants modifies nanoparticle surface properties, enabling the design of tailored nanoparticle-based additives for foam stabilization or destabilization in industrial applications. Similarly, Yang and Liu (2010) article demonstrates the versatility of surface-functionalized nanoparticles in enhancing dispersion stability and preventing aggregation, with notable applications in thermal engineering processes involving phase-change heat transfer. Together, these works underscore the potential of functionalized nanoparticles in diverse fields, from advanced heat transfer technologies to precise foam control in industrial systems.
- **Nanofluids for Superior Thermal Management**: Nanofluids—fluids containing dispersed nanoparticles—exhibit enhanced thermal conductivity and heat capacity, which improve heat dissipation in EDUs. The improved thermal management properties help reduce the formation of foam and microbubbles by minimizing localized temperature gradients that can cause gas to come out of solution. Krajnik et al. (2011) states that nanofluids can achieve up to a 40% increase in thermal conductivity at optimal nanoparticle concentrations while maintaining fluid stability. This leads to better cooling performance, which minimizes thermal deformation of materials, reduces friction at tool-workpiece interfaces, and decreases wear on cutting tools. Furthermore, improved convective heat transfer rates allow for precise temperature control in processes like die casting and grinding, resulting in higher product quality and extended equipment lifespan.

1.10.2 Innovative Testing and Diagnostic Tools

The development of advanced testing and diagnostic tools is critical for understanding and mitigating foaming and aeration in real-time. Innovations in sensor technology and data analytics are enabling more precise measurements and predictive maintenance capabilities, providing engineers with greater insight into fluid dynamics within EDUs.

1.10.2.1 Developments in Real-Time Monitoring of Foaming and Aeration

Traditional laboratory-based testing methods, while valuable for baseline assessments, often fall short when it comes to monitoring foaming and aeration in real-time, especially under dynamic operating conditions. To address these limitations, new in-situ measurement technologies are being developed, allowing for precise, localized monitoring without altering the system design.

- **Miniaturized Embedded Sensors**: Recent advancements have led to the development of miniaturized sensors that can be embedded directly into critical areas of the EDU lubrication and cooling systems. These sensors, which include optical, acoustic, and micro-electromechanical systems (MEMS)-based designs, are capable of measuring localized aeration levels and foam characteristics in real-time. Due to their small size and minimal impact on fluid flow, these sensors provide high-fidelity data without disrupting the system's normal operation. This technology enables precise monitoring of bubble dynamics and foam stability in areas such as gear meshes, bearing surfaces, and heat exchangers. Wang et al. (2022) introduced a novel sensor network designed for real-time, in situ monitoring of wear debris in lubricating oil. This system employs a coaxial capacitive sensor network with parallel curved and non-parallel plane electrodes, enabling the detection of changes in oil permittivity caused by the presence of wear particles. The design allows for integration directly into oil pipelines, facilitating continuous monitoring of both metallic and non-metallic debris, which is crucial for assessing machinery health and preventing potential failures.
- **Non-Invasive Measurement Techniques**: Non-invasive methods, such as ultrasonic and magnetic resonance sensors, can measure gas content and fluid dynamics without direct contact. These methods are particularly useful in opaque fluids where traditional optical methods are not feasible. By leveraging sound wave or magnetic field interactions with the fluid, these sensors can provide a detailed profile of air content and bubble size distribution throughout the system. The article by Hauptmann et al. (1998) reviews the current advancements and emerging trends in ultrasonic sensors for process monitoring and chemical analysis, highlighting their versatility and precision in industrial and laboratory applications. These sensors utilize ultrasonic waves to monitor parameters such as fluid density, viscosity, and concentration, enabling real-time and non-invasive measurements. The authors discuss state-of-the-art developments, including miniaturization, enhanced signal processing, and integration with IoT systems, which improve sensor performance and broaden their application scope. The article underscores the growing role of ultrasonic sensors in optimizing processes, ensuring quality control, and advancing chemical analysis technologies.
- **Integration with Existing Control Systems**: Advanced sensor technologies can be integrated with existing control systems, enabling automated adjustments to operating parameters based on real-time data. This integration allows for active aeration and foam management, adjusting flow

rates, pressures, or fluid compositions as needed to maintain optimal performance. The 2022 editorial "Integration of Sensors in Control and Automation Systems 2020" highlights advancements in sensor technologies, such as intelligent sensors and distributed networks, that enhance fault detection, precision control, and system efficiency. It emphasizes the critical role of sensors in modernizing industries like automation and robotics, driving improvements in process optimization and reliability.

The emergence of these novel sensing technologies enables engineers to accelerate the development of state-of-the-art solutions for EDU design, offering unmatched precision and flexibility in monitoring and managing fluid behavior under varying conditions.

1.10.2.2 Integration of AI and Machine Learning in Predictive Maintenance

Artificial intelligence (AI) and machine learning (ML) are revolutionizing fluid diagnostics and predictive maintenance strategies. These technologies enable the processing and analysis of large datasets collected from advanced sensors, providing predictive insights and identifying potential issues before they impact performance.

- **AI-Based Predictive Models**: Machine learning algorithms can process historical data on fluid behavior to identify patterns and predict foaming and aeration events. These models are trained to recognize the onset of aeration based on variables such as pressure, temperature, and flow rate, allowing for preemptive intervention. For example, Agarwal et al. (2019) discuss the use of ensemble-based machine learning to enhance foam control in industrial fermenters, a critical challenge in bioprocess management. By leveraging multiple machine learning models, the approach improves the detection and prediction of foaming conditions, enabling timely and precise interventions to mitigate foam formation. The authors emphasize how this advanced method contributes to more efficient foam management, reducing the reliance on antifoam agents and enhancing the sustainability of industrial fermentation processes.
- **Digital Twins for Real-Time Simulation**: Digital twins—virtual replicas of physical systems—are being used to simulate fluid behavior within EDUs under various operating conditions. By integrating real-time sensor data with digital twin models, engineers can continuously monitor system performance, identify potential issues, and implement corrective actions with minimal downtime. The 2021 paper by Thomas et al. (2021) presents a CFD digital twin designed to analyze and optimize miscible fluid blending processes, offering real-time simulations to understand flow dynamics, mixing efficiency, and operational performance. This tool provides valuable insights into blending behaviors, enabling improvements in process design and operational strategies. The digital twin aims to enhance efficiency and sustainability in fluid mixing across various industrial applications.
- **Automated Test Systems**: AI-driven automated test systems can dynamically adjust testing parameters based on sensor feedback. Luo et al. (2024)

explores the development of explainable and generalizable AI-driven multiscale informatics frameworks for modeling dynamic systems. By integrating AI techniques with domain knowledge, the approach enhances the transparency, interpretability, and adaptability of models across various scales and complex systems. This method enables a comprehensive evaluation of fluid performance under complex transient conditions by accurately capturing and analyzing dynamic interactions, providing valuable insights for optimizing systems in fields such as energy, healthcare, and manufacturing. These systems can simulate a wide range of operating scenarios, from start-stop cycles to high-load endurance tests, providing deeper insights into fluid stability and performance.

1.10.3 Sustainable and Eco-Friendly Solutions

Sustainability and environmental responsibility are increasingly becoming key considerations in the selection and development of lubricants and coolants for EVs. Future research is focusing on biodegradable additives, eco-friendly formulations, and life-cycle assessments to ensure that the next generation of EV fluids meets both performance and environmental standards.

1.10.3.1 Biodegradable Additives and Fluids

The development of biodegradable additives and fluid formulations is gaining momentum as manufacturers seek to minimize the environmental impact of their products. These additives offer the dual benefits of high performance and reduced ecological footprint.

- **Plant-Based Esters and Synthetic Alternatives**: Plant-based esters are being explored as replacements for conventional mineral oils. These esters provide excellent lubrication and thermal properties while being fully biodegradable and non-toxic. Synthetic alternatives, such as polyalkylene glycol (PAG)-based fluids, are also being developed to meet stringent environmental and performance requirements. For example, Kin Yu Lam from Cargill Bioindustrial discusses the advantages of natural esters over synthetic alternatives as dielectric fluids in his article (Lam, n.d.), focusing on their environmental benefits and superior performance characteristics. He emphasizes that natural esters, derived from renewable sources, offer enhanced biodegradability, lower toxicity, and better fire safety compared to synthetic esters. The article concludes by highlighting the increasing importance of sustainable and efficient dielectric fluids in addressing environmental concerns and improving electrical equipment reliability.
- **Biodegradable Surfactants and Defoamers**: Research into biodegradable surfactants and defoamers is focusing on natural sources such as fatty acids and amino acid derivatives. These components are not only effective at controlling foam formation but also break down quickly in the environment, reducing long-term ecological impact. The article *A Review on Natural Surfactants* (De et al., 2015) highlights the sustainability, biodegradability,

and reduced toxicity of surfactants derived from renewable sources such as plants and microbes, emphasizing their broad potential in industries like cosmetics, pharmaceuticals, and food. Similarly, the article *New Generation of "Green" Defoamers for Challenging Drilling and Cementing Applications* (Bava et al., 2013) presents eco-friendly defoamers that effectively mitigate foaming issues while adhering to strict environmental standards in oil and gas operations. Both articles underscore the importance of environmentally conscious innovations, with the former focusing on the versatility of natural surfactants and the latter on targeted advancements in green defoaming technologies.

1.10.3.2 Environmental Considerations in Fluid Selection and Testing

Environmental considerations are becoming a critical factor in the selection and testing of lubricants and coolants for EDUs. Engineers are increasingly focusing on life-cycle assessments, evaluating the environmental impact of fluids from production to disposal.

- **Reduction of Harmful Substances**: Formulating fluids without harmful substances such as heavy metals or persistent organic pollutants (POPs) is a priority. Research is exploring alternative chemistries that maintain high performance while eliminating environmentally harmful components. For example, the development of a sustainable metalworking fluid (MWF) formulated from natural, biodegradable, and non-toxic components is proposed by Lodhi and Kumar (2021). The MWF comprises soybean oil as the dispersed phase, gum acacia as a natural emulsifier, and water as the continuous phase, with additives like neem oil and guar gum enhancing its properties. Performance evaluations during grinding operations indicate that this eco-friendly MWF outperforms traditional mineral oil-based fluids, offering reduced grinding forces, improved surface finishes, and better surface morphology. The study underscores the potential of such natural ingredient-based MWFs to replace conventional fluids, thereby mitigating environmental and health hazards associated with mineral oil-based products.
- **Sustainable Manufacturing Practices**: Implementing sustainable manufacturing processes, such as using renewable energy and minimizing waste, is becoming standard practice in fluid production. These practices not only reduce the environmental impact but also align with global sustainability goals and regulations. In their review, Pimenov et al. (2022) analyze strategies for resource conservation and environmental sustainability in manufacturing through optimization techniques and advanced machining environments. They highlight the potential of methods like energy-efficient machining, tool path optimization, and eco-friendly cutting fluids to reduce waste and energy consumption, promoting sustainable practices across industries. The authors also identify future opportunities in integrating digital technologies and life cycle analysis to further enhance sustainability in manufacturing processes.

1.11 CONCLUSION

1.11.1 Summary of Key Insights on Foaming and Aeration in EDUs

The investigation into foaming and aeration in Electric Drive Units (EDUs) has revealed the significant impact these phenomena have on the thermal management, lubrication efficiency, and overall reliability of electric vehicle (EV) powertrains. Foaming and aeration manifest in various forms, such as entrained air bubbles, surface foam, and dissolved gases, each of which can detrimentally affect fluid dynamics and system performance.

One of the primary concerns is the reduction in lubrication film thickness caused by entrained air, leading to increased friction and wear. Similarly, the presence of foam within the fluid can degrade the heat transfer properties, causing localized hot spots and thermal inefficiencies. These issues are exacerbated under high-speed, high-load conditions typical of EDUs, where rapid pressure and temperature fluctuations can lead to the outgassing of dissolved gases and subsequent bubble formation.

In terms of fluid design, the role of additives such as anti-foaming agents and surfactants is crucial in managing foam stability and mitigating the effects of aeration. Advanced formulations, including nanotechnology-enhanced lubricants, have shown promise in reducing foaming propensity and improving overall system efficiency. Simulation and experimental methods have been instrumental in characterizing foam behavior and developing strategies to optimize fluid performance.

1.11.2 Final Thoughts on the Importance of Addressing These Challenges

The implications of foaming and aeration extend far beyond mere fluid dynamics. These phenomena can compromise the durability of critical components, increase energy consumption, and degrade the performance of thermal management systems. Addressing these challenges is not only essential for ensuring the longevity and efficiency of EV powertrains but also for meeting the evolving standards of reliability and sustainability in the automotive industry.

Given the complexity of modern EV systems, it is no longer sufficient to rely on traditional testing methods. Advanced in-situ sensors, real-time monitoring, and integration of AI-driven predictive maintenance tools are being explored for their abilities in managing foaming and aeration effectively. Engineers must consider these factors during the early design stages of new EV platforms, employing a holistic approach that incorporates both fluid chemistry and system architecture to mitigate these issues.

1.11.3 Implications for Future EDU Design and Reliability

The findings on foaming and aeration emphasize the need for a proactive approach in future EDU designs. From a design perspective, incorporating features such as air-bleed valves, optimized flow channels, and degassing systems will play a crucial role in preventing the introduction and accumulation of air in the fluid circuits.

Additionally, fluid formulations need to be tailored to the specific operating conditions of EDUs, with particular focus on minimizing air solubility and maximizing thermal conductivity.

The adoption of innovative testing methodologies, including real-time sensors and advanced CFD models, allows for better prediction and management of foaming and aeration under dynamic conditions. This will enable engineers to identify potential problem areas and implement corrective measures before issues arise. Moreover, the use of biodegradable and eco-friendly additives that maintain performance while reducing environmental impact aligns with the broader sustainability goals of the EV industry.

In conclusion, understanding and managing foaming and aeration is critical for advancing the reliability and efficiency of electric drive units. Through the integration of advanced fluid formulations, optimized system designs, and cutting-edge diagnostic tools, the next generation of EVs can achieve superior performance and set new benchmarks for durability and energy efficiency in the automotive sector.

ACKNOWLEDGEMENTS

We would like to thank PMT Tamson Instruments and Chongqing Gold Mechanical & Electrical Equipment Co., Ltd for granting us permission to use their product images in this book.

PMT PM Tamson Instruments

GOLD
格纳迪机电
master@hy-industry.com

REFERENCES AND FURTHER READING

Agarwal, A., Liu, Y. A., & McDowell, C. (2019). 110th anniversary: Ensemble-based machine learning for industrial fermenter classification and foaming control. *Industrial & Engineering Chemistry Research*, *58*(36), 16719–16729. https://doi.org/10.1021/acs.iecr.9b02424.

Baran, B. A. (2007). *Engine Lubrication Oil Aeration*. Thesis, Massachusetts Institute of Technology, Massachusetts.

Bava, L., Mahmoudkhani, A., Wilson, R., & Levy, L. (2013). New generation of "Green" defoamers for challenging drilling and cementing applications. In: *SPE Production and Operations Symposium,* Oklahoma City, Oklahoma, USA. https://doi.org/10.2118/164504-ms.

Cadix, A., Meeker, S., Kaushik, S., Haumesser, E., & Ovarlez, G. (2021). Associative microgels, new self adaptive systems to control fluid loss in well cementing. *Day 3 Wed,* November 17, 2021. https://doi.org/10.2118/207472-ms.

Canter, N. (2015, December). *Foam: Dealing with a Persistent Problem*. Society of Tribologists and Lubrication Engineers. https://www.stle.org/files/TLTArchives/2015/12_December/Feature.aspx.

Chandran Suja, V. (2022). Challenges in mitigating lubricant foaming. *Lubricants*, *10*(6), 108. https://doi.org/10.3390/lubricants10060108.

Chokelarb, W., Sriprom, P., Permana, L., & Assawasaengrat, P. (2024). Assessment of overall remaining useful life of lubricants by integrating oil quality and performance. *Heliyon, 10*(18). https://doi.org/10.1016/j.heliyon.2024.e37486.

Chongqing Gold Mechanical & Electrical Equipment Co., Ltd. (2025). *ASTM D3427 Automated Air Release Properties Analyzer of Hydrocarbon Based Oils [Image].* Chongqing Gold Mechanical & Electrical Equipment Co., Ltd. https://chinainstrument.en.made-in-china.com/product/nNOQptfJsBVk/China-ASTM-D3427-Automated-Air-Release-Properties-Analyzer-of-Hydrocarbon-Based-Oilshtml?tradeFrom=3_3.

Chun, S. M. (2008). Aeration effects on the performance of a turbocharger journal bearing. *Tribology International, 41*(4), 296–306. https://doi.org/10.1016/j.triboint.2007.08.006.

De, S., Malik, S., Ghosh, A., Saha, R., & Saha, B. (2015). A review on natural surfactants. *RSC Advances, 5*(81), 65757–65767. https://doi.org/10.1039/c5ra11101c.

Goodwin, M. J., Dong, D., Yu, H., & Nikolajsen, J. L. (2007). Theoretical and experimental investigation of the effect of oil aeration on the load-carrying capacity of a hydrodynamic journal bearing. *Proceedings of the Institution of Mechanical Engineers, Part J: Journal of Engineering Tribology, 221*(7), 779–786. https://doi.org/10.1243/13506501jet279.

Hauptmann, P., Lucklum, R., Püttmer, A., & Henning, B. (1998). Ultrasonic sensors for process monitoring and chemical analysis: State-of-the-art and trends. *Sensors and Actuators A: Physical, 67*(1–3), 32–48. https://doi.org/10.1016/s0924-4247(97)01725-1.

Hurtado, Y., López, D., Lopera, S. H., Cortés, F. B., & Franco, C. A. (2021). Influence of surfactant adsorption on surface-functionalized silica nanoparticles for gas foam stability. In: Mishra, B. (Ed.), *Lecture Notes in Nanoscale Science and Technology*, Springer, New York, pp. 339–357. https://doi.org/10.1007/978-3-319-12051-5_9.

Kimura, Y., Kato, N., Kataoka, T., Takahashi, K., & Kikuchi, T. (2011). Bubble behavior in engine lubricant. *International Journal of Automotive Engineering*, *2*(4), 149–153. https://doi.org/10.20485/jsaeijae.2.4_149.

Krajnik, P., Pusavec, F., & Rashid, A. (2011). Nanofluids: Properties, applications and sustainability aspects in materials processing technologies. In: Hesselbach, J. and Herrmann, C. (Eds.), *Advances in Sustainable Manufacturing*, Springer, New York, pp. 107–113. https://doi.org/10.1007/978-3-642-20183-7_16.

Lam, K. Y. (n.d.). A tale of two esters – Why the natural one is the better dielectric fluid. In: *IEEE Transmission & Distribution Conference & Exhibition*, Dallas, Texas, USA, May 21–24, 2006. https://www.cargill.com/doc/1432242852210/a-tale-of-two-esters-why-the-natural-one-is-better.pdf.

Li, S., Lin, X., & Jiang, S. (2023). A lubrication model for bubbly oil lubrication with interfacial effect and thermal effect using the theory of multiphase mixtures. *Friction, 12*(2), 271–290. https://doi.org/10.1007/s40544-023-0756-4.

Lodhi, A. P., & Kumar, D. (2021). Natural ingredients based environmental friendly metalworking fluid with superior lubricity. *Colloids and Surfaces A: Physicochemical and Engineering Aspects, 613*, 126071. https://doi.org/10.1016/j.colsurfa.2020.126071.

Luo, C., Li, A.-J., Xiao, J., Li, M., & Li, Y. (2024). Explainable and generalizable AI-driven multiscale informatics for dynamic system modelling. *Scientific Reports, 14*(1). https://doi.org/10.1038/s41598-024-67259-4.

Manz, D., & Cheng, W. (2006). On-line measurements of engine oil aeration by X-ray absorption. In: *ASME 2006 Internal Combustion Engine Division Spring Technical Conference (ICES2006)*, Aachen, Germany, pp. 719–726. https://doi.org/10.1115/ices2006-1356.

Mastrone, M. N., Concli, F., & Concli, F. (2023). Numerical modeling of fluid's aeration: Analysis of the power losses and lubricant distribution in gearboxes. *Journal of Applied and Computational Mechanics, 9*(1), 83–94. https://doi.org/10.22055/jacm.2022.40666.3625.

Morales, R., Fernández-Caballero, A., Somolinos, J. A., & Sira-Ramírez, H. (2022). Integration of sensors in control and automation systems 2020. *Journal of Sensors*, 2022, 1–3. https://doi.org/10.1155/2022/9765679.

Naif. (2024). *GPT Image Generator [AI-Generated Image].* https://chatgpt.com/-image-generator.

Noria Corporation. (2013, February 8). *Preventing Micropitting and Surface Fatigue.* Machinery Lubrication. https://www.machinerylubrication.com/Read/29276/micropitting-surface-fatigue?utm_source=chatgpt.com.

Obiols, J. (2003b). Lubricant oxidation monitoring using FTIR analysis - application to the development of a laboratory bulk oxidation test and to in-service oil evaluation. In: *SAE Technical Paper Series: 2003 JSAE/SAE International Spring Fuels and Lubricants Meeting*, Yokohama, Japan. https://doi.org/10.4271/2003-01-1996.

Pimenov, D. Y., Mia, M., Gupta, M. K., Machado, Á. R., Pintaude, G., Unune, D. R., Khanna, N., Khan, A. M., Tomaz, Í., Wojciechowski, S., & Kuntoğlu, M. (2022). Resource saving by optimization and machining environments for Sustainable manufacturing: A review and future prospects. *Renewable and Sustainable Energy Reviews*, *166*, 112660. https://doi.org/10.1016/j.rser.2022.112660.

PM Tamson Instruments. (2024). *Tamson ASTM D892 Foaming Dual Bath [Image].* PM Tamson Instrumentshttps://tamson-instruments.com/tamson-astm-d892-foaming-dual-bath-230v-50-60hz.

Prolić, T. Ć., & Lepušić, A. (2012). Effect of foaming on the antiwear properties of lubricating oils. *Fuels and Lubricants: Journal for Tribology, Lubrication, Application of Liquid and Gaseous Fuels and Combustion Engineering*, *51*(1), 38–46.

Ren, C., Zhang, X., Jia, M., Ma, C., Li, J., Shi, M., & Niu, Y. (2023a). Antifoaming agent for lubricating oil: Preparation, mechanism and application. *Molecules*, *28*(7), 3152. https://doi.org/10.3390/molecules28073152.

Sander, R. (2023). Compilation of Henry's law constants (version 5.0.0) for water as solvent. *Atmospheric Chemistry and Physics*, *23*(19), 10901–12440. https://doi.org/10.5194/acp-23-10901-2023.

Sattar, M. A., Naser, J., & Brooks, G. (2013). Numerical simulation of creaming and foam formation in aerated liquid with population balance modeling. *Chemical Engineering Science*, *94*, 69–78. https://doi.org/10.1016/j.ces.2013.01.064.

Savant Labs. (2024, August 22). *Air Separation: The Key Performance Indicator.* https://www.savantlab.com/testing-highlights/air-separation-astm-d3427-air-release-testing/#:~:text=Why%20is%20ASTM%20D3427%20Air,and%20how%20long%20they%20last.

Shahnazar, S., Bagheri, S., & Abd Hamid, S. B. (2016). Enhancing lubricant properties by nanoparticle additives. *International Journal of Hydrogen Energy*, *41*(4), 3153–3170. https://doi.org/10.1016/j.ijhydene.2015.12.040.

Shore, J. F., Christodoulias, A. I., Kolekar, A. S., Lockwood, F. E., & Kadiric, A. (2022). *Prediction of Electric Vehicle Transmission Efficiency Using a New Thermally Coupled Lubrication Model.* SAE Technical Paper Series. SAE International, Warrendale, Pennsylvania, USA. https://doi.org/10.4271/2022-01-5026.

Tao, M., Slike, J., Bhagat, M., Srinivasan, C., Zhang, Y., & Motin, A. (2022). Numerical modelling of coolant filling and de-aeration in a battery electric vehicle cooling system. *SAE International Journal of Advances and Current Practices in Mobility*, *5*(1), 75–83. https://doi.org/10.4271/2022-01-0775.

Thomas, J., Sinha, K., Shivkumar, G., Cao, L., Funck, M., Shang, S., & Nere, N. K. (2021). A CFD digital twin to understand miscible fluid blending. *AAPS PharmSciTech*, *22*(3). https://doi.org/10.1208/s12249-021-01972-5.

Triple R Filtration. (2024, August 16). *The Impact of Sludge on Hydraulic and Lubricating Systems: Oil Filtration*. https://triplerfiltration.com.au/impact-of-sludge-on-hydraulic-and-lubricating-systems/

Vavra, C. (2022, August 31). *Air Entrainment, Aeration, Cavitation and Foaming: How Are they Related*? Plant Engineering. https://www.plantengineering.com/articles/air-entrainment-aeration-cavitation-and-foaming-how-are-they-related/.

Wang, J., Liu, L., Zhang, S., Liao, B., Zhao, K., Li, Y., Xu, J., & Chen, L. (2023). Review of the perspectives and study of thermo-responsive polymer gels and applications in oil-based drilling fluids. *Gels*, *9*(12), 969. https://doi.org/10.3390/gels9120969.

Wang, Y., Lin, T., Wu, D., Zhu, L., Qing, X., & Xue, W. (2022). A new in situ coaxial capacitive sensor network for debris monitoring of lubricating oil. *Sensors*, *22*(5), 1777. https://doi.org/10.3390/s22051777.

Yang, X., & Liu, Z. (2010). A kind of nanofluid consisting of surface-functionalized nanoparticles. *Nanoscale Research Letters*, *5*(8), 1324–1328. https://doi.org/10.1007/s11671-010-9646-6.

Yano, A., Watanabe, S., Miyazaki, Y., Tsuchiya, M., & Yamamoto, Y. (2004). Study on sludge formation during the oxidation process of turbine oils. *Tribology Transactions*, 47(1), 111–122. https://doi.org/10.1080/05698190490278985.

2 Solving for Corrosion and Conductive Deposit Concerns

Gregory Miiller

2.1 IDENTIFYING THE PROBLEM

Product failure due to corrosion is estimated to cost $1 trillion annually [1]. It is the industry's responsibility to discover, monitor, resolve and implement resolutions. This responsibility is especially critical when applications are in highly corrosive environments where product failure not only is extremely costly, but also raises the risk of catastrophe and/or human harm. There is reason to believe that many professionals, although aware of general facts specific to corrosion, do not maintain adequate knowledge of how and why diverse methodologies for corrosion prevention.

Let's first define the word corrosion, to better understand cause and effect. The National Association of Corrosion Engineers (NACE): Corrosion is the deterioration of a substance, usually a metal, or its properties, because of an undesirable reaction with its environment. Corrosion is a natural and inevitable process that, once understood, can be mitigated so that preventive measures and controlled outages can take place [2].

Consideration of the environmental conditions in which the electrical components will be placed. Conditions such as moisture, dust, and temperature can affect the rate of corrosion. The level of corrosion typically increases with moisture content. Common atmospheric sources of moisture are rain, dew, and condensation. Material particles, at a granular or smaller level such as soot, can cling to surfaces and even promote or proliferate the retention of moisture. Increasing the temperature of a corrosive environment will generally increase the rate of corrosion. As standard chemistry implies, for every 10°C rise in the temperature, the corrosion rate can potentially double.

Materials that commonly corrode the most are metals. There are certain types of corrosion that can be identified. Here are just a few of the types of corrosion that may be experienced in these cases. General corrosion, the most basic type, is typically caused by a chemical reaction that results in the deterioration of the entire exposed surface of a metal in a uniform manner. Ultimately, the metal deteriorates to the point of failure. There is also galvanic corrosion that may occur between two dissimilar metals. If these metals are placed in contact (or otherwise electrically connected), this potential difference produces electron flow between them, causing corrosion. Then there is crevice corrosion, which is a localized corrosion, that is

DOI: 10.1201/9781003496069-2

associated with a stagnant solution located in material gaps, flaws, seals, joints, and other inconsistent surface areas. Pitting is a form of corrosion caused by a localized attack resulting in holes in the metal. Erosion corrosion results when a protective layer of oxide on a metal surface is dissolved or removed by a source such as pressure from a fluid, particulate, or even entrained air, exposing the underlying metal to further corrode and deteriorate. Other types of corrosion are not discussed as the application here does not apply.

In the automotive industry, there are several combinations of materials that can corrode. One material of concern is copper. Copper corrodes when exposed to a variety of aggressive agents including acids, salts, phosphates, and sulfur compounds. When the environment is a formulated hydrocarbon solution, such as fuels and lubricants, the main concern is corrosion by sulfur compounds. In systems where there are moving parts under load, such as automotive applications, the lubricant formulation will include additives that contain sulfur. These are necessary for the protection of ferrous gears and bearings but potentially detrimental to copper. It is important that the lubricant is tested to assure it is appropriate to the application; see the connector failure in Figure 2.1.

Conventional vehicles, those without an electric motor to propel or assist propulsion, contain several lubricants such as motor oil, transmission fluid, and axle oil. In these vehicles, the copper components exposed to the lubricant are present mainly in the form of mechanical hardware, and while these lubricants must maintain a set level of corrosion protection, this characteristic is not a primary concern. In electrified vehicles, lubricants also contact copper components that are present in the form of electric motor windings, wiring, and busbars. Here, corrosion can result in catastrophic failure, and the ability of the lubricant to prevent corrosion is a key concern. Electrified vehicles generally operate at lower temperatures

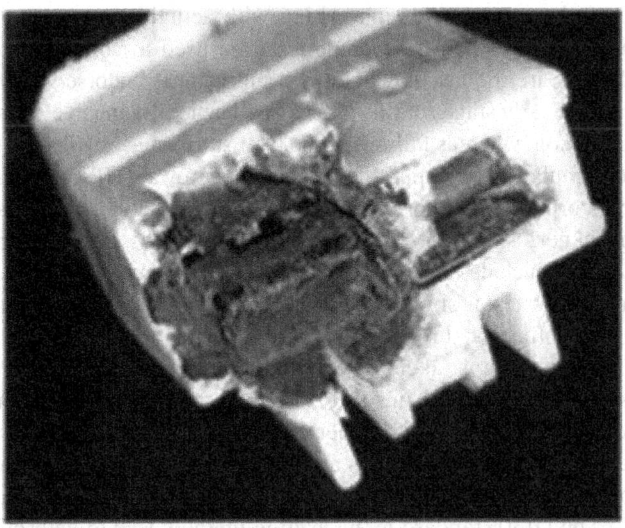

FIGURE 2.1 Corrosion on EV connector.

than their conventional counterparts, but with excursions to high temperatures at the electric motor windings. Mechanical parts can weaken and then fail due to corrosion-related material loss. Similarly, material loss of electrical components can lead to failure by creating open or intermittent circuits. However, the products of corrosion can also deposit on circuits or within electric motor windings. Copper sulfides are semiconductors, and such accumulations can lead to shorts that may impact circuit behavior or burn out the electric motor. It is therefore necessary to truly understand how corrosion varies with temperature from ambient to the maximum local temperature within electrified drives and the potential impact of accumulated corrosion products.

2.2 AUTOMOTIVE INDUSTRY CORROSION TESTS

There are several bench tests that have been developed over the years to determine a fluid's performance as it relates to corrosion. Typically, these tests are relatively inexpensive and representative of the application conditions. Instrumentation of these types evaluates both the durability of the lubricant and the compatibility with the hardware with a smaller investment versus engine testing.

Some tests utilize copper and other metal materials, albeit solid or soluble, as catalysts to react within the testing regime. One example is the ASTM test D2722 for engines and turbine fluids. A copper coil is utilized to facilitate the oxidation process. Other tests are similar, such as the High Temperature Corrosion Bench Test D6594 and the Thin Film Oxidation Update Test D4742. Both tests utilize other materials such as tin and lead. The test D6594 has coupons that are solid, whereas D4742 contains soluble metals that are in a catalyst solution.

The ASTM D130 standard is the most common bench test used to assess the potential for copper corrosion [3]. Other bench tests, more specific to the application, may then be used, such as corrosion testing with actual copper-containing parts. Longer term and more expensive module tests are used to confirm proper performance. Field trials are the last stage before a lubricant is endorsed for market use. Failures during the early stages of this process will generally involve alteration of the lubricant formulation, whereas failures at the later stages may involve adjustments to the lubricant formulation or minor changes to the hardware. Bench tests described here are also incorporated in lubricant specifications as they provide a rapid assessment of lubricant performance. Most lubricant specifications, including transmission and axle lubricants, include the ASTM D130 or its equivalent.

The premise of the test is simplistic in nature. A clean copper strip was immersed in a sample of gasoline for three hours at 50°C. At the conclusion of the test, any corrosion or discoloration is taken to indicate the presence of corrosive sulfur. The end-of-test strips are ranked using the ASTM Copper Strip Corrosion Standard Adjunct, a colored reproduction of copper strips characteristic of various degrees of tarnish and corrosion. The strip results are divided into four classifications describing the various appearances under each classification that match letter codes shown in Figure 2.2. The D130 standard is the leading method used to determine the corrosiveness of various fuels, lubricants, and other hydrocarbon-based solutions to copper [3].

Rating	Designation	Description
1a	Slight Tarnish	Light orange, nearly the same as freshly polished strip
1b		Dark orange
2a	Moderate Tarnish	Claret red
2b		Lavender
2c		Multicolored with lavender blue or silver, or both, overlaid on claret red
2d		Silvery
2e		Brassy or gold
3a	Dark Tarnish	Magenta overcast on brassy strip
3b		Multicolored with red and green showing (peacock), but no gray
4a	Corrosion	Transparent black, dark gray or brown with peacock green barely showing
4b		Graphite or flat black
4c		Glossy or jet black

FIGURE 2.2 D130 rankings [3].

2.3 NEW TEST DEVELOPMENTS FOR ELECTRIC VEHICLES

As one can imagine, a subjective test such as D130 can be problematic for the industry. This is especially true in cases where shades and colors change due to the chemistry of the fluid. Ranking becomes more difficult when red, orange, and tan produce results that are not decipherable when they are dark or light in comparison to one another.

Vehicle technology is rapidly changing, due mainly to the electrification of drivetrains. A greater number of electric circuits, used for sensors and electro-mechanical devices, are positioned such that they are in contact with the drivetrain lubricant or lubricant vapor. If the lubricant is not properly formulated, this can lead to corrosion of the electric contacts. These corrosion products are conductive in nature, and the damage created on the circuits can result in anything from minor faults to catastrophic failure. In hybrid electric and electric vehicles (HEVs and EVs), when the electric motor is in contact with the lubricant or lubricant vapor, corrosion within the motor windings can cause a direct short resulting in motor burn-out.

Subsequently, the D130 extended test began to be developed. The main reason is that the exposer time and temperatures do not match electric vehicle conditions [4]. The concept was multifaceted in nature. Not only are the times and temperature modified, but at the end of the test the amount of soluble copper residing in the resulting sample is measured. The technique is to utilize the inductively coupled plasma apparatus, ICP, to comparatively measure the copper parts per million against a known standard. Unfortunately, this test does not address surface corrosion that can build up on the surface, causing inconsistent ICP copper measurements.

However, these rankings can be valuable if a metals analysis of the lubricant is conducted after the test, such as the new Wire Corrosion Test (WCT). However, this does not provide insight into the risks presented by the formation of conductive deposits. A new corrosion and deposit test that can predict the propensity of conductive deposit formation is the conductive deposit test (CDT). Both tests are quickly becoming test methods and inducted into specifications. The details of the tests and their intrinsic value are important to understand.

2.4 THE WIRE CORROSION TEST

Original equipment manufacturers (OEMs) have turned to electrification to maintain regulatory compliance enacted to reduce greenhouse gases and limit global warming. Consequently, a greater number of electric circuits, used for sensors and electro-mechanical devices, are incorporated into the drivetrain and could be positioned such that they are in contact with the lubricant or lubricant vapor. Corrosion of electronic components raises more significant concerns than corrosion of mechanical parts, particularly if the corrosion occurs within the windings of an electric motor.

Exposed copper in circuits and motors is susceptible to corrosion mainly by sulfur species contained in the lubricant, usually as additives to protect the gears and bearings but also from the byproducts of the base oil refinement. To prevent these additives from corroding copper, specific inhibitors are included in the lubricant. The need for gear and bearing protection while preventing the corrosion of copper has always been a challenging aspect of lubricant formulation and is critical. Consideration for these properties for lubricants is important for the next generation of HEVs and EVs [3]. This has led to the development of the WCT.

The concept of this test method is not new, however. In 1960, a similar test was developed to measure the corrosion rate of a steel wire within lubricant additives [5]. Resistance measurements are made throughout the test and were applied to antiwear and extreme pressure additives. The system was resistively heated and cooled back down. The actual diameter was also measured at the end of the test. Since that time, various temperatures and times have been analyzed to best duplicate the conditions of the environment within the engine [6,7].

The WCT was further developed to better simulate the areas of an electric vehicle where the motor, connectors, and other electrical components are immersed in the oil. The design intentionally assesses both solution and vapor phase corrosivity simultaneously, using a copper wire submerged in a test lubricant and a second copper wire suspended in the headspace above, as shown in Figure 2.3. Copper not in direct contact with the lubricant can suffer corrosive damage through contact with the vapor in the unit headspace. In several scenarios, vapor phase corrosion is more severe than that of the solution phase. A low current, approximately 1 mA, is applied through the wires, and resistance is measured as a function of time. Electrical resistance can be measured with extreme precision and accuracy, and subsequently this

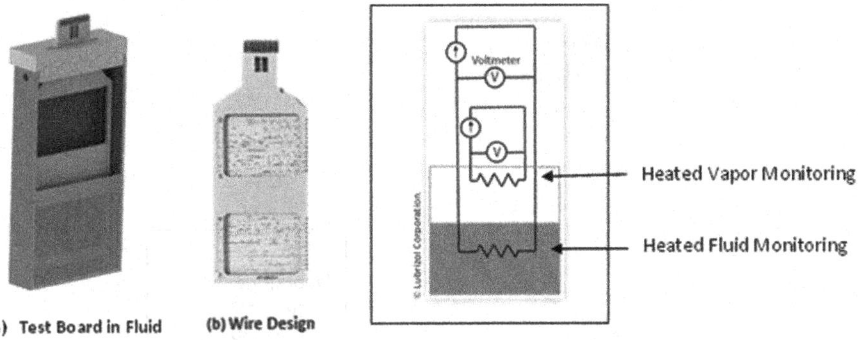

FIGURE 2.3 WCT testing layout.

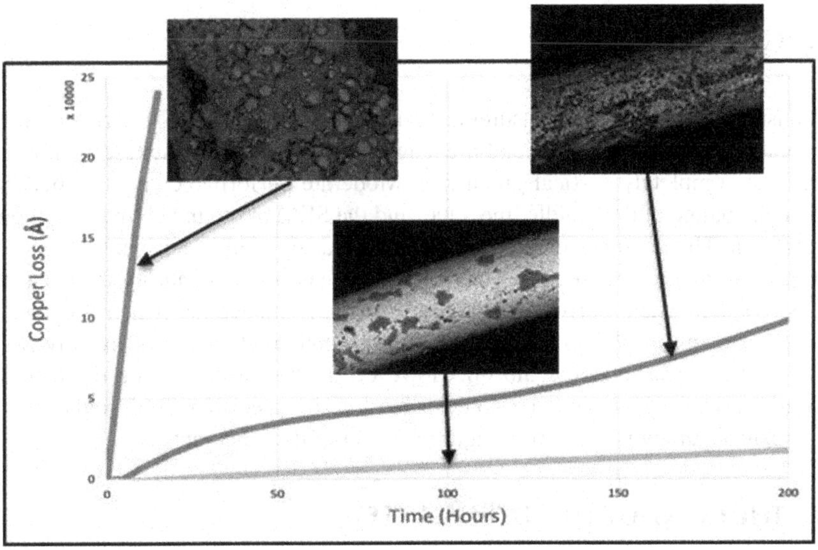

FIGURE 2.4 SEM wire view.

test method is highly sensitive, which enables the detection of copper corrosion rapidly at operationally applicable temperatures. The rate of change figures was initially evaluated in units of resistance change per time but are converted to an equivalent loss of wire radius per time to express the corrosion rate in more conventional units.

The premise of the test is to measure the rate of corrosion that takes place over time. Specific conditions, to best predict EV drivetrain data, are set to test at two different temperatures. Other temperatures can be measured, but 130°C and 150°C have been determined as most critical for performance evaluation. The time duration selected for the average rate of change, in angstroms, is 72 hours.

After many experiments and subsequent analysis on the scanning electron microscope, SEM, the corrosion amount and rate of change measured were established. In other words, the SEM showed that the diameter of the wire was directly proportional to the measurement of angstroms/hour in the actual WCT test. Figure 2.4 shows the

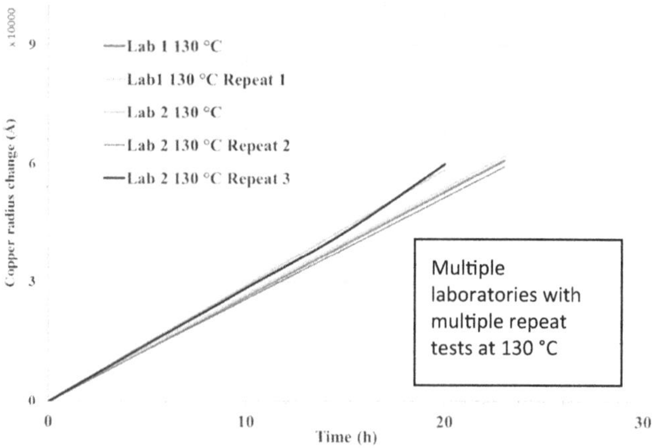

FIGURE 2.5 Repeatability at 130°C.

relationship between those measurements and the subsequent SEM corrosion measurements. The extreme failure and breakage of the wire during the test is shown by the almost completely vertical green line. Moderate performance is clear between the rate of change of the middle line, blue, and the SEM photo indicating the amount of corrosion. The good performing fluid is shown as the almost horizontal line.

The most remarkable measure of this testing has been the ability to repeat the results. Oxidation and corrosion tests are inherently inadequate when it comes to precise repeatability. This is known universally. Some oxidation tests are only 70%–75% accurate on a holistic scale [8]. However, in this case, the rate of corrosion change, angstroms/hour, is quite noteworthy. Interim precision would indicate that the percent accuracy is over 90%. Figure 2.5 exemplifies that fact.

2.5 THE CONDUCTIVE DEPOSIT TEST

As previously indicated, the WCT represents new technology to measure and quantify corrosion. It does not, however, assess the damage that corrosion products can cause to electronics or electric motors. Most lubricants contain active sulfur, which may corrode copper and lead to the formation of copper sulfides. These corrosion products form dendrites and deposits that may be conductive and can accumulate between electrical contacts. This seemingly small issue may result in circuit errors and electrical shorts. In the case of e-motor windings, this can lead to motor burn out [9] as shown in Figure 2.6 (example of arced motor windings). A method to quantify the propensity to form conductive deposits, a conductive deposit test (CDT), is needed.

It has been researched that the corrosion depletion of the WCT and conductive deposits, the CDT, have different responses [10]. Additionally, the variations of copper corrosion tests do not truly predict either of these scenarios. So, encapsulating the entire corrosion and deposit occurrences would require both tests to paint the entire picture.

Bridging deposits require time to generate, and initial versions of these tests typically use a duration of 1000 hours and a temperature of 150°C. This duration is

FIGURE 2.6 Failed motor windings.

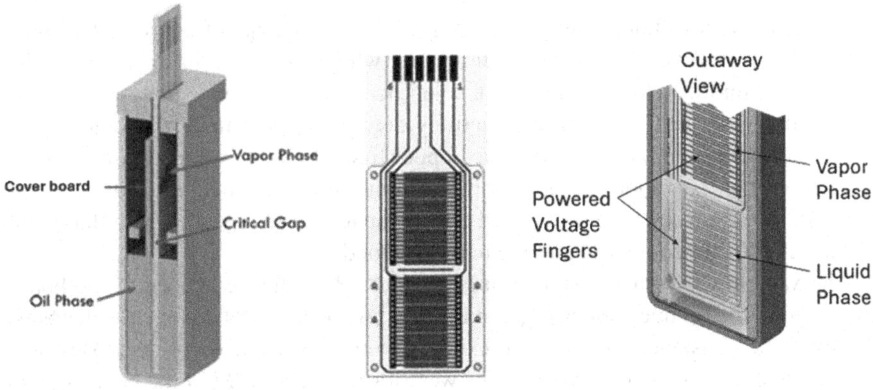

FIGURE 2.7 CDT testing layout.

required in part due to the spacing distance between the non-connected copper paths. The temperature of 150°C accelerates the corrosion process and dendrite formation. In setups with one PCB, the board is generally energized with 5 V DC. In setups involving multiple boards, at least one board will be energized.

A commercial version of the CDT was developed and made into an ASTM test method D8544. The test cell, in this case, incorporates an optimized PCB with a cover board. The gap between the boards is patented and determined as critical in the formation of the deposits as compared to EV motors and other electronics. Automotive companies confirmed that small gaps in connectors, small cavities in the drivetrain, and spaces between electronics were concerning. Therefore, a cover board with a precision offset is placed over the test circuit board to simulate these gaps. The test cell and PCB design are depicted in Figure 2.7. Because of the extended duration

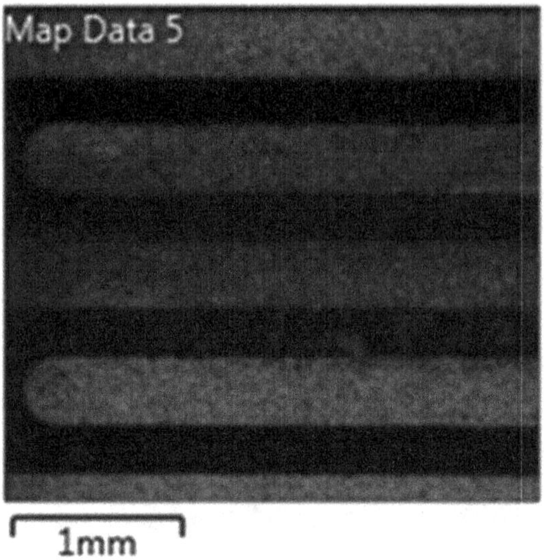

FIGURE 2.8 EDS copper map of PCB with dendrites.

of the CDT, the test fluids may oxidize, forming a wide range of byproducts. Many of these byproducts can be corrosive in nature, which can result in oxidation of the copper trace and lead to the removal of solubilized copper, thus removing traces on the circuit board. At that point, conductivity between copper traces becomes nullified and inaccurate. Thus, a conductive copper trace, known as the continuity loop, is monitored in the vapor and liquid phases to monitor the proclivity of this occurrence. If the continuity loop is depleted during the test, then any data from that point forward is considered invalid and the test is stopped.

An example CDT data is shown in Figure 2.8. Here, the resistance of the board immersed in the oil does not change, indicating no formation of conductive deposits. However, the resistance of the board exposed to the lubricant vapor decreases over time, indicating that conductive deposits were formed. The SEM of the board shown in Figure 2.8 shows the conductive dendrites forming.

Interpreting when this resistance loss indicates a problematic level of conductive deposits is not straightforward. This is particularly difficult when the resistance loss is transient. In the example shown in Figure 2.9, the resistance decreases slightly at ~100 hours before increasing to the baseline, and again more significantly at ~200 hours, returning once more to the baseline, before continually decreasing at ~250 hours until the value plateaus at ~700 hours. A simple resistance loss limit may not be adequate to characterize a lubricant's propensity to form conductive deposits or to serve as a comparison characteristic to distinguish lubricants and rank their performance. A method that accounts for the resistance loss and the resistance fluctuations is needed.

In principle, an integration of the resistance loss curve with time would account for both the permanent resistance loss and the transient events. This summation of resistance loss with time can be appropriately scaled to create a conductive deposit

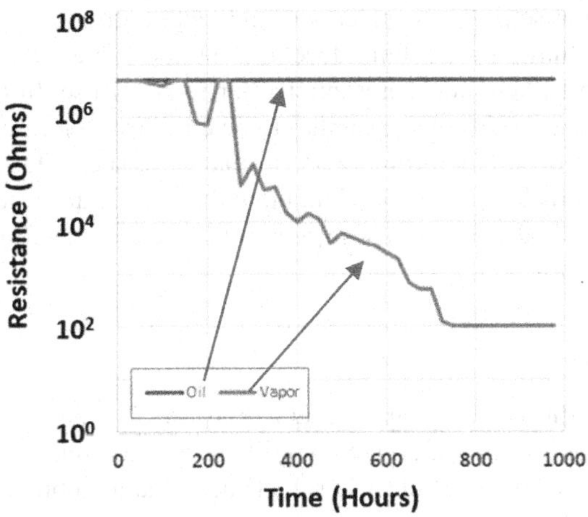

FIGURE 2.9 Schematic of example CDT data shown on a logarithmic scale.

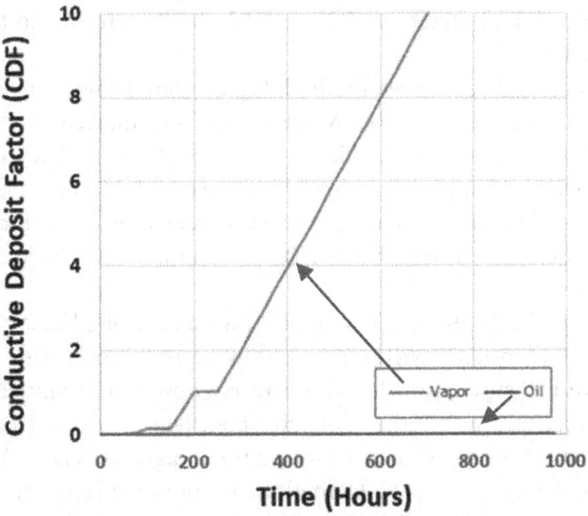

FIGURE 2.10 CDF transformation of CDT data shown in Figure 2.8.

factor (CDF). The value of this factor would then represent the severity of deposit formation and could be used to set a time criterion for lubricant comparisons. Expressing the CDT data shown in Figure 2.9 in terms of this CDF summation transformation gives the curves shown in Figure 2.10. The transformation of the solution phase data, of course, is trivial (zero summation). The vapor phase transformation simplifies the information into a linear curve, where horizontal movements are indicative of the contribution associated with transient events. This curve is smooth and easy to interpret. If the criterion limit for acceptable performance is set as CDF = 2, the hours to this criterion for the vapor phase is 300.

The specification-setting organizations and automotive suppliers utilize this information to set limits for the fluid development companies. The values here are aggressive and most likely not representative for industry use. In other words, it is more feasible to view a CDF of 4 or higher to be a failure, versus one of 2 or lower. The test was also reduced to 500 hours. Performance of such a test longer than that period is typically breaking down the fluid more than necessary when compared to a sealed electric vehicle drivetrain. These conditions are considered when developing international standards.

2.6 SUMMARY

Corrosion is a condition that plagues every industry. The automotive industry, in particular, has spent millions of dollars to resolve corrosion-related issues within the engine and drivetrain. With the electric vehicle age of automobiles, the problem has not subsided. Therefore, a new era of testing is needed to assist the industry.

Current testing, such as the ASTM D130 copper strip test, is a long-established technique to measure corrosion. Many others also exist. However, because of the differences in conditions and requirements with electric vehicles, new technologies have been developed, and more will follow. Both the drivetrain and battery cooling areas are being addressed.

The WCT and CDT are new methodologies that provide new insights not gleaned from traditional strip tests. While comparing lubricants to each other is possible at any set of conditions, relating test results to real-world applications requires a knowledge of applications and how various lubricants perform in them and in these tests. Currently, there are several 'research laboratory' versions of these tests being conducted within the industry, and the differences between them are not established.

Two new test methods, the WCT and CDT, were described that provide a better gauge of how well a lubricant will protect hardware from corrosion. The WCT provides quantitative measurements that characterize how well a lubricant can protect the dissolution of hardware in both the oil and vapor phases over a wide range of temperatures. The test is fast, highly sensitive, and unambiguous. The CDT can quantify the propensity of conductive deposit formation in a lubricated environment, a definite concern in applications such as HEVs and EVs where increasingly electronics and electric motors are in contact with lubricant.

Both the WCT and CDT are useful tools in developing lubricants suitable for electric drivetrains. The examples of applying the CDT to determine the impact of ester base oil and a traction reduction additive demonstrate information that is directly applicable to advancing lubricant formulation for electrified powertrains.

Adoption of the WCT and CDT by organizations such as ASTM will provide stronger tools to the automotive industry and benefit the development of electrified drivetrains [11]. Both tests measure different aspects of the corrosion process and together provide a complete profile to ascertain the hardware protection of lubricants.

REFERENCES

1. Ellis, S., "Avoiding corrosion in electrical systems properly specifying electrical products for highly corrosive environments will reduce overall long-term cost and risk of failure", *Consulting Engineer Magazine*, Texas, February 17, 2014.
2. Linxiao, Z., "Electrochemical characterization of an oil/water alternately wetted rotating cylinder electrode", *Corrosion Journal of Science and Engineering*, 77(1), 72–84, 2021.
3. Hunt, G., 100 Years of corrosion testing – is it time to move beyond the ASTM D130? The wire corrosion and conductive layer deposit tests", *SAE International Journal of Fuels and Lubricants, 17*(1), 17–35, 2024.
4. Aikin, A. R., "Evolution of copper corrosion testing for electric vehicle lubricants", *Tribology & Lubrication Technology*, 80(1), 2024. https://www.stle.org/files/TLTArchives/2024/01_January/Feature.aspx
5. Barcroft, F. T., "A technique for investigating reactions between E.P. additives and metal surfaces at high temperatures", *Wear*, 3(6), 440–453, 1960.
6. Warren, B., Hunt, G., Bryant, M., Neville, A., Morina, A. and Gahagan, M., "Automatic transmission fluid corrosion inhibitor interactions with copper", *Lubrication Science*, 30, 301–315, 2018.
7. Hunt, G., "New perspectives on the temperature dependence of lubricant additives on copper corrosion", *SAE International Journal of Fuels and Lubricants*, 10(2), 521–527, 2017.
8. ASTM Test Method D2272, *The Standard Test method for Oxidation Stability of Steam Turbine Oils by Rotating Pressure Vessel*, Version May 2022, ASTM D02 Petroleum Products.
9. Hunt, G. and Prengaman, C., *Understanding Vapor and Solution Phase Corrosion of Lubricants Used in Electrified Transmissions*. SAE Technical Paper 2020-01-0561, SAE International, Warrendale, PA, 2020.
10. Miiller, G., Pelz, R., Mueller, G. and Gillespie, D., Frontiers manuscript, Thesis, Saginaw Valley State University, Saginaw, 2024,
11. Wang, Y. and Zhou, X., "Comprehensive design and optimization of an electric vehicle powertrain equipped with a two-speed dual-clutch transmission", *Advances in Mechanical Engineering, 9*(1), 1–13, 2017.

3 Electrical Properties of EV Fluids

Berthold Bode, Simon Hausner, André Heine, and Maik Schröder

3.1 INTRODUCTION

For well-known reasons, electric mobility has become a major subject of global research in recent years. As voltages in electric vehicles (EV) are by far higher than in conventional internal combustion engine vehicles (ICEV), electrical stress is an important cause of wear in the electric drivetrain: spark-erosive discharges may occur at small gaps – such as found in bearings – when there is a difference in the electrical potentials involved, and these breakdowns can irreversibly damage both the operating fluids and the surrounding mechanical components.

As a consequence, special EV fluids are being developed: lubricants, gear oils, immersion coolants, etc., that combine thermophysical and electrical properties, permitting them to withstand the exposure to electric fields. While it is important to electrically characterize fluids for effective damage prevention, their dielectric properties can also serve as reliable indicators of the fluid condition, e.g., when monitoring oil aging or analyzing the lubrication condition. Also looking at the traditional standards for dielectric testing of insulating liquids, this chapter is to give you an introduction to the relevant electrical properties of EV fluids, focusing on moderately conductive lubricants. Newly created test methods are described that analyze these fluids with respect to their operating conditions. In order to do so, we follow an extensive approach and gradually establish real-life conditions. First, we conduct temperature-controlled tests based on a constant electrode gap under atmospheric pressure. We then expand the static analysis with pressurized tests. Finally, we use a tribometer with a fully variable gap to dynamically analyze the electrical properties to their full extent.

3.2 DETERMINING THE DIELECTRIC PROPERTIES

In any application where electrical currents are present, knowing the dielectric properties of the utilized EV fluids is mandatory for ensuring the desired running conditions. These properties are:

- the *specific electrical conductivity* and the *specific electrical resistivity*,
- the *relative permittivity* and
- the *dielectric dissipation factor*.

 DOI: 10.1201/9781003496069-3

After their modes of operation are explained, an accurate test method for the dielectric properties is presented that will help with the development of robust and sustainable EV fluids.

3.2.1 Specific Electrical Conductivity and Resistivity

An important electrical characteristic of any EV fluid is its ability to transport electrical charges, typically given as the specific electrical conductivity κ (or sometimes σ). This ability is bound to the molecular components of the fluid. To understand the process of electrical conduction, we will look at two planar metal plates with an electrolyte solution, e.g., sodium chloride (NaCl) dissolved in water, between them, as shown in Figure 3.1.

When diluted in water, the salt components Na^+ and Cl^- separate from each other and move freely as charged particles in the solution. As long as there is zero voltage applied to the metal plates, the ions move randomly through the fluid, resulting in no measurable transport of electrical charge. As soon as a voltage is applied to the metal plates, the positively charged sodium ions (Na^+) will start to move towards the negatively charged electrode (anode), while the negatively charged chlorine ions (Cl^-) move towards the positively charged electrode (cathode), resulting in a measurable electrical current between the metal plates. Evidently the electrical current in the fluid depends on the molecular particles present in the fluid. The more charged particles there are and the higher their mobility, the stronger the resulting electrical conductance. The specific electrical conductivity κ gives an indication of this behavior. The higher the electrical conductivity is, the more electrical current can pass through the fluid (given a constant voltage). Since most commonly used oils tend to be electrically insulating – meaning they have low electrical conductivities below 1 µS/m – the reciprocal value of the electrical conductivity is often used for reference. It is called specific *electrical resistivity* ρ:

$$\rho = \frac{1}{\kappa} \tag{3.1}$$

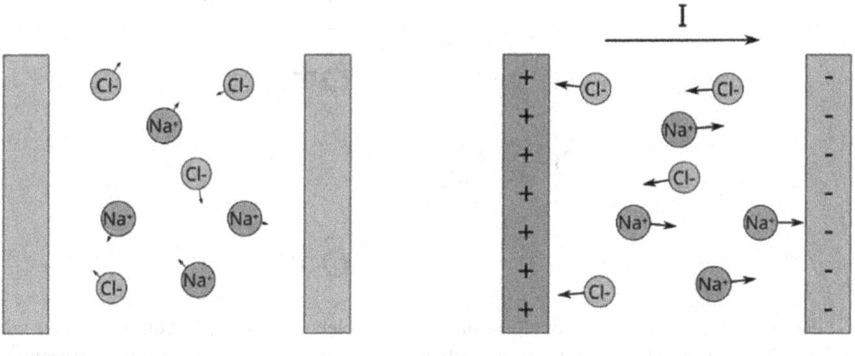

FIGURE 3.1 Salt solvent between two metal plates. *Left*: free ion movement, no net current. *Right*: charge-dependent ion movement with measurable net current *I* between the plates.

In classical insulating lubricants, the charged particles come from impurities like dust, humidity, or broken molecular bonds due to thermal aging. Bringing in a small number of charged particles in an otherwise non-conducting material greatly increases the electrical conductivity and hence drastically reduces the electrical resistivity. This high sensitivity makes both of these properties useful indicators for the purity and condition of a fluid.

3.2.2 Relative Permittivity

Aside from its ability to transport electrical charges by physically moving the charged particles it contains, the fluid can also change the alignment of its particles inside an electric field. This quality is indicated by the *relative permittivity* ϵ_r, also known as the *dielectric constant.*

A water molecule, for example, has an asymmetrical structure, as shown in Figure 3.2: both of its hydrogen atoms (H) lean towards one side. Due to the inherent high tendency of oxygen (O) to attract electrons, the electrical bonds tend to pull the electrons from the hydrogen atoms closer to the oxygen atom.

This leads to an asymmetrical charge distribution within the molecule, where the oxygen side appears to be negatively charged in contrast to the hydrogen side, giving the molecule a so-called *electrical dipole moment.* Such asymmetries are very common in molecular structures; hence, the resulting effect of applied fields should be taken into account when dealing with electric applications.

Let us again imagine two metal plates with water between them. With zero voltage applied to the plates, the water dipoles are randomly oriented with no electric field measurable. When applying a voltage, the dipoles rotate to align themselves according to the electric field E_{ext} imposed by the charged metal plates. Looking at the aligned dipoles, one can see that their electric field E_{dipole} actually opposes the external field (see Figure 3.3). Furthermore, the electric field created by the aligned fluid molecules creates an attractive force pulling the charges closer to the fluid. This results in more charges being pulled towards the dielectric. Hence, the number of charges in the plates rises, depending on the strength of the dipoles. This means that the higher the electric field from the dielectric molecules, the more charges can be

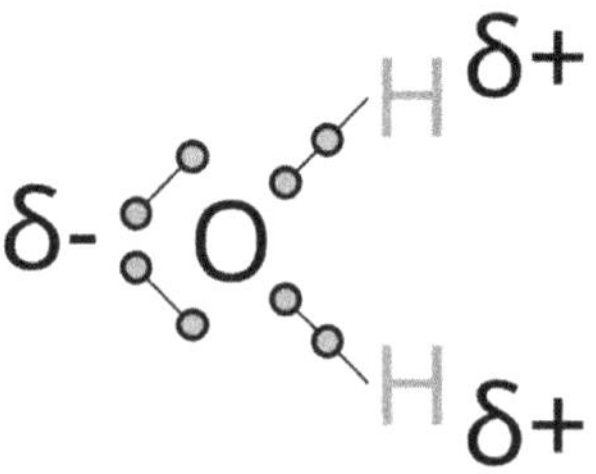

FIGURE 3.2 Schematic representation of a water molecule. The hydrogen atoms (H) are bonded to the oxygen atom (O) through electron bridges (a pair set of electrons, represented as dots). Due to the higher electronegativity of the O atom, the H electrons are pulled closer, resulting in an apparent negative charge on the oxygen side and an apparent positive charge on the hydrogen side of the molecule.

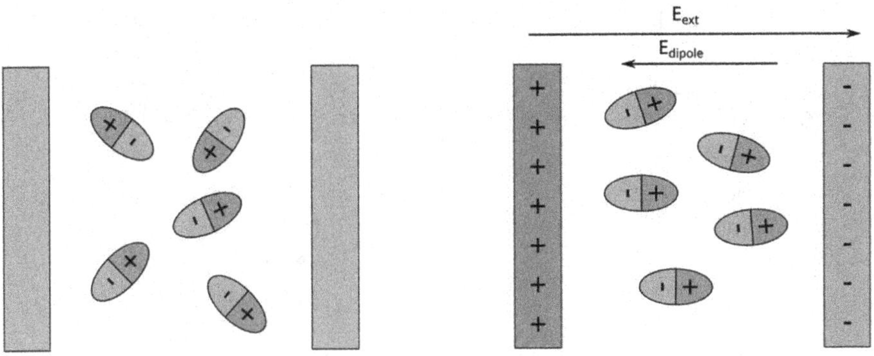

FIGURE 3.3 Electrically asymmetric molecules between two metal plates. *Left*: Random molecule orientation without any external electric field. *Right*: Charge-dependent molecule orientation with opposing electric fields.

stored in the metal plates. The metal plates can be seen as a capacitor, for which the capacitance C defines the amount of charges that can be stored at a given voltage. The capacitance can be calculated as

$$C = \frac{\epsilon_0 \epsilon_r A}{d} \tag{3.2}$$

The relative permittivity ϵ_r represents the strength of the resulting opposing electric field in the dielectric. It is a substance-specific property that is highly temperature-dependent. The constant ϵ_0 is the permittivity of the vacuum. A is the area of the plates (electrodes), while d equals the distance between the plates.

While non-polar molecules like hydrocarbons don't have inherent (permanent) dipole moments, the external field induces a small shift in the molecular charge distributions, creating temporary dipoles aligned with the external field, again creating an opposing electrical field.

3.2.3 Dielectric Dissipation Factor

Another important characteristic of the dielectric is given by the *dielectric dissipation factor*, also known as *tan δ* or *DDF*. To understand this property, one can again imagine a simple plate capacitor model. If the space between the plates is perfectly insulating, the current I_c will lead the voltage by exactly 90° since there is no current flow in phase with the voltage. In reality, the dielectric separating the plates will always allow for a minuscule current I_R between the plates to flow in phase with the voltage. This constitutes an additional current path in parallel to the ideal capacitor, leading to a power loss. The vectorial representation (Figure 3.4) shows that tan δ equals the ratio of I_R and I_C:

$$\tan \delta = \frac{I_R}{I_C} \tag{3.3}$$

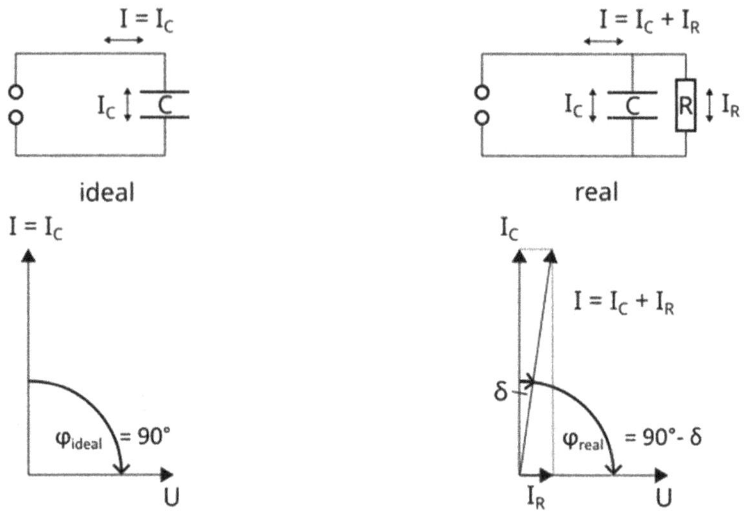

FIGURE 3.4 Equivalent circuit diagram and vectorial representation of the current I and the voltage U of a capacitor. *Left*: In an ideal capacitor the current I leads the voltage U by exactly 90°. *Right*: In a real capacitor the dielectric between the electrodes will always show some form of conductance, allowing a current I_R in phase with the voltage U. This leads to a phase shift of the combined current I to be less than 90°. The deviation is defined as the loss angle δ.

Thus, tan δ gives information about the energy dissipation inside the material, commonly appearing in the form of heat losses when storing electrical energy in the capacitor.

3.2.4 Dielectric Tests with the Pioneering DIN 51 111 Method

When it comes to determining the dielectric properties of fluids, various standards have been utilized over the years. An overview of the most common methods is given in Table 3.1.

All these standards focus entirely on insulating liquids (pS/m range). Dealing with EV fluids, one often faces the challenge of significantly higher electrical conductivities ($\kappa > 1$ nS/m), either due to their specific formulation or due to contamination and/or aging. With the ongoing development of more electrically specialized substances, a new standard had to be established to determine their dielectric characteristics.

The *DIN 51 111* standard "Electrical properties of fresh and used oils from electric drives in vehicles - Measurement of the specific electrical conductivity, the relative permittivity (ε_r) and the dielectric dissipation factor (tan δ)" [1] was published in 2024. Being designed specifically for the high-precision determination of the dielectric fluid properties, it introduces a non-damaging procedure of electrical lubricant testing. This section is to give a quick overview over this pioneering method.

A sectional model of the cylindrical capacitor test cell of flucon's *EPSILON+* Dielectricity Meter which was the sole instrument used for the *DIN 51 111*

TABLE 3.1
Commonly Used Standards for Determination of the Dielectric Properties of Fluids

Designation	Title
ASTM D924	Standard test method for dissipation factor (or power factor) and relative permittivity (dielectric constant) of electrical insulating liquids
ASTM D1169	Standard test method for specific resistance (resistivity) of electrical insulating liquids
ASTM D2624	Standard test methods for electrical conductivity of aviation and distillate fuels
ASTM D4308	Standard test method for electrical conductivity of liquid hydrocarbons by precision meter
DIN 51412	Testing of petroleum products – determination of the electrical conductivity
IEC 60247	Insulating liquids – measurement of relative permittivity, dielectric dissipation factor (tan δ) and D.C. resistivity
IEC 61620	Insulating liquids – determination of the dielectric dissipation factor by measurement of the conductance and capacitance – test method

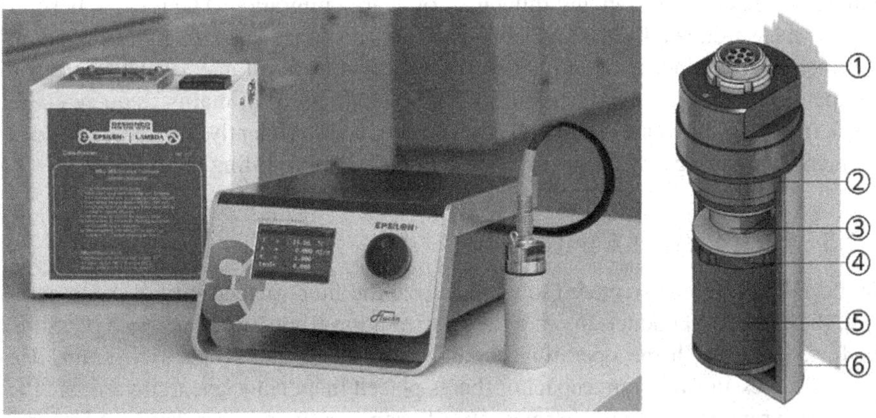

FIGURE 3.5 EPSILON+ dielectricity meter (complies with DIN 51 111) and a sectional model of its test cell: ① sensor connector, ② outer electrode (screw-on cup with thread), ③ ground, ④ electrical insulation, ⑤ inner electrode (with integrated Pt100 temperature sensor), and ⑥ Insulating cup [2].

standardization is shown in Figure 3.5. Its electrodes have an area of 30 cm^2 and a gap of 1 mm. The required sample size is 8 mL per test. The test ranges are 20 pS/m to 600 µS/m for the electrical conductivity (equaling 1.67 kΩm to 50 GΩm for the resistivity), 1–10 for the permittivity and 0–10,000 for the dissipation factor.

To account for the high sensitivity of the dielectric properties to temperature changes, *DIN 51 111* requires isothermal conditions with a ±0.1°C tolerance. The method recommends 50°C, 100°C, and 150°C as standard test temperatures for oils. These isotherms were chosen to include a potentially moisture-influenced

TABLE 3.2
Reference Values for Relative Permittivity and Specific Electrical Conductivity Given by DIN 51 111

Substance	Air	*n*-Decane ($C_{10}H_{22}$)		
Temperature	Room temperature	50°C	100°C	150°C
Relative permittivity ε_r	1.000 ± 0.009	1.948 ± 0.025	1.880 ± 0.025	1.810 ± 0.025
Specific electrical conductivity κ	N/A	<0.2 nS/m	<0.2 nS/m	<0.4 nS/m

measurement (50°C), a moisture-unaffected measurement (150°C), and a reference point for typical operating temperatures (100°C). Further round-robin tests have been carried out recently within the DGMK *NA 062–06–53-A* project to extend the DIN method for grease testing. In this case, 50°C, 80°C, and 100°C are the designated three test isotherms.

At each temperature, the specific electrical conductivity is measured with a low AC voltage between 5 and 30 Hz, preferably at 20 Hz. This will prevent electrode damage by electrolysis and any influences of charge migration. The relative permittivity is measured at a frequency between 10 and 100 kHz. The *DIN 51 111* standard also includes reference data for plausibility testing, shown in Table 3.2. The dielectric dissipation factor is to be specified for 50 Hz AC (European mains frequency). To check if the highly sensitive test cell has been cleaned sufficiently, it is recommended to check the reading with air inside (1.000 ± 0.009) before filling in a sample.

3.2.5 High-Pressure Capacitor Tests

While *DIN 51 111* was created to characterize the fluid dielectricity over temperature, the electrical conductivity and the relative permittivity also vary with pressure. To further approach the operating conditions given in the electric drivetrain, one can use a pressure-resistant capacitor that is placed in an autoclave to investigate the dielectric properties in an isothermal setup. flucon uses an in-house developed assembly that allows for automated, temperature-controlled tests of various fluid properties, such as viscosity, thermal conductivity, compressibility, velocity of sound, and dielectricity, up to pressures of 1.4 GPa (14,000 bar) (see Figure 3.6).

The resulting data is especially valuable for tribological model calculations of the elastohydrodynamic lubrication (EHL) contacts of bearings and gears, e.g., when using the *pressure-viscosity coefficient* α_p for Hamrock and Dowson's equation to determine the *film thickness* $h_{\min}$ [3]. When raising the pressure, the electrical conductivity decreases, whereas the permittivity typically increases. The following results were found for dielectric tests with FVA 3A reference mineral oil on the one hand and polyglycol base oil with and without a small dose (0.05 m%) of highly conductive ionic liquid added to it on the other. As can be seen below, the permittivity of polyglycol is in general significantly higher than that of mineral oil. This needs to be taken into account when evaluating impedance results, and in particular the measured capacitance, as presented in the next section (Figures 3.7 and 3.8).

FIGURE 3.6 Fluid analyses in flucon's automated high-pressure fluid laboratory.

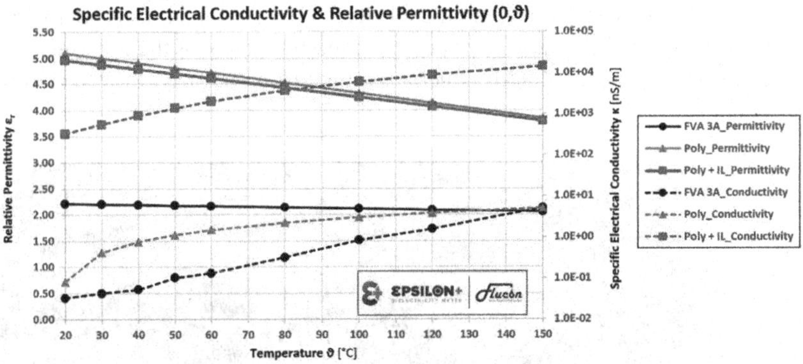

FIGURE 3.7 Dielectric properties at atmospheric pressure (DIN 51 111 test with additional increments).

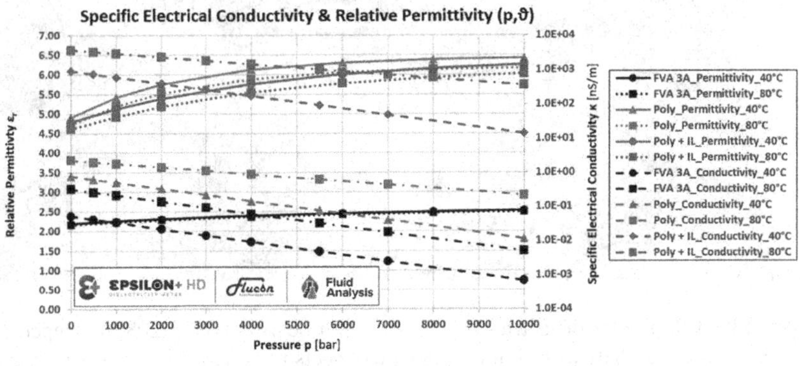

FIGURE 3.8 Dielectric properties at high pressures.

3.2.6 Bearing Impedance Measurement

Just as the dielectric properties of lubricants are increasingly becoming the focus of development work, more and more tribometers are being electrified and new test equipment is being developed. In many of these applications, the aim is to estimate or at best measure, the behavior of the lubricant in the tribological contact under the influence of an electric field. One convenient approach is to use a roller bearing test rig that is equipped with current and voltage measuring technology.

Flucon is following this approach with the development of the *E-Lub Tester* (Figure 3.9), a dielectric tribometer based on the "GESA" set-up developed at the MEGT of the University of Kaiserslautern-Landau [4]. The tribometer axially loads and drives a horizontally arranged thrust ball bearing of type 51208 in a temperature-controlled measuring insert that is filled with a lubricant sample. The measuring insert is galvanically and capacitively insulated from the drive unit. While the rotating upper bearing ring is connected through a sliding contact, the stationary lower ring is contacted through the bottom of the bearing cup, which also includes channels for precise temperature control. The bearing cup is sealed without contact, so that an electrical current can only flow through the lubricant, and hence through the lubricating film in the tribo-contact (Figure 3.10).

When the typical elastohydrodynamic lubricant (EHL) film builds up in the rotating roller bearing, its dimensions can be estimated by means of the load-dependent Hertzian area and the minimum film thickness at point contact according to Hamrock and Dowson.

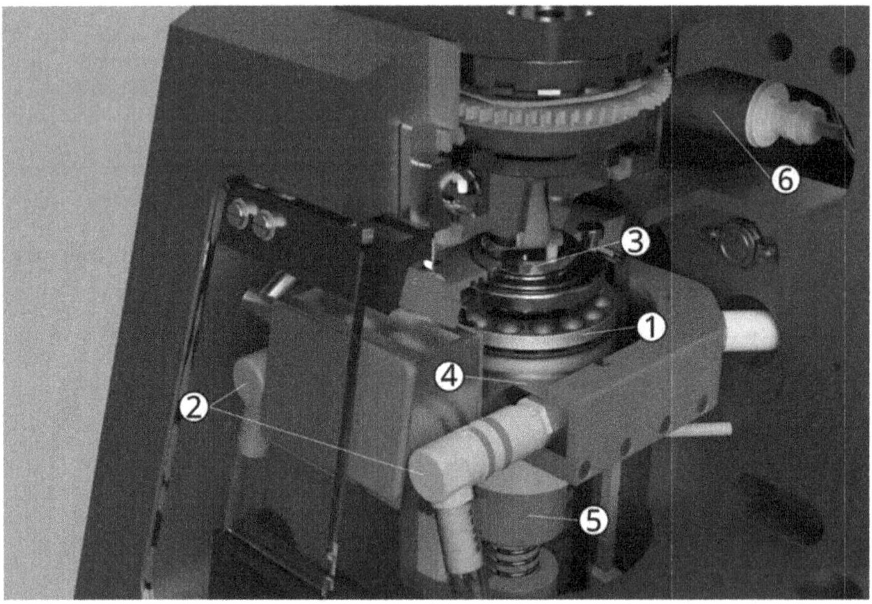

FIGURE 3.9 E-lub tester dielectric tribometer: ① test bearing type 51208, ② temperature control connectors, ③ shaft with sliding contact (connects rotating bearing ring), ④ bearing cup (connects stationary bearing ring), ⑤ axial force actuator, and ⑥ load cell [5].

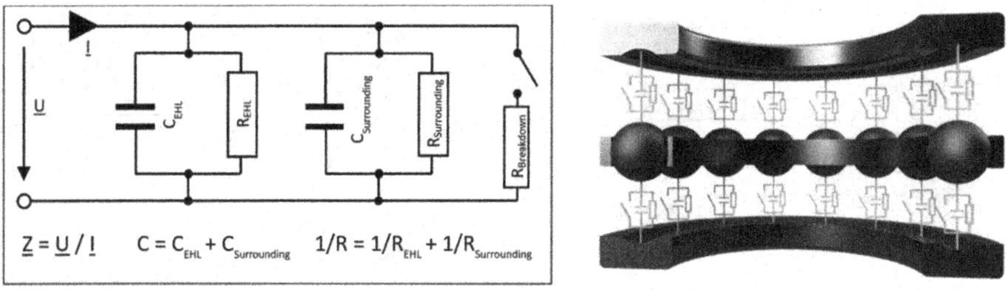

FIGURE 3.10 Electrical equivalent circuit diagram and 3D sectional model of the roller bearing with lubricant film impedances.

The EHL contact can be simplified as a capacitor, with the lubricant as a dielectric. Depending on the electrical conductivity of the lubricant, a charge exchange occurs – the lubricant therefore forms an ohmic resistance connected in parallel to the capacitance ($C_{\mathrm{EHL}} \| R_{\mathrm{EHL}}$, Figure 3.10). In addition to the EHL contact, the area adjacent to it must be considered as well ($C_{\mathrm{Surrounding}} \| R_{\mathrm{Surrounding}}$) [6]. Although this area takes up more space than the EHL contact, due to the different curvatures of the rolling element and the raceway, surface distances much larger than the usual 100–600 nm given in the EHL contact are effective. Since small electrode distances affect the capacitance the most, $\mathrm{C}_{\mathrm{EHL}}$ makes up the major part of the overall capacitance C.

If you now carry out impedance spectroscopy by applying a frequency-variable alternating voltage and by measuring the current, you will obtain the typical amplitude and phase responses of the complex impedance Z of R–C parallel circuits, which corresponds to the chosen electrical equivalent circuit. Figure 3.11 illustrates this relationship and at the same time suggests deriving the impedance parameters from the measured impedance magnitude $|Z|$ using suitable regression methods.

For correct impedance measurement on a moving bearing, the relatively small lubrication gap size must be taken into account to prevent exceeding the critical field strength which would result in an electrical breakdown. The measuring voltage should be designed accordingly or should be adjustable.

3.2.7 Why Measure the Impedance?

Impedance spectroscopy provides two parameters from which conclusions can be drawn about the lubrication condition of the bearing at freely definable operating points and about the condition of the lubricant itself.

To recall the aforementioned theory (Section 3.2.1): the C value depends, in principle, on how all the molecules in the gap are aligned in the alternating electric field. The R value, on the other hand, depends on how many free charge carriers are present in the gap and how mobile they are. These charge carriers can also be provided by a relatively small amount of molecules. Based on this consideration, it is also logical that the C value is rather insensitive to influences of e.g., contamination, lubricant aging, or humidity, while the R value indicates changes in the lubricant very quickly.

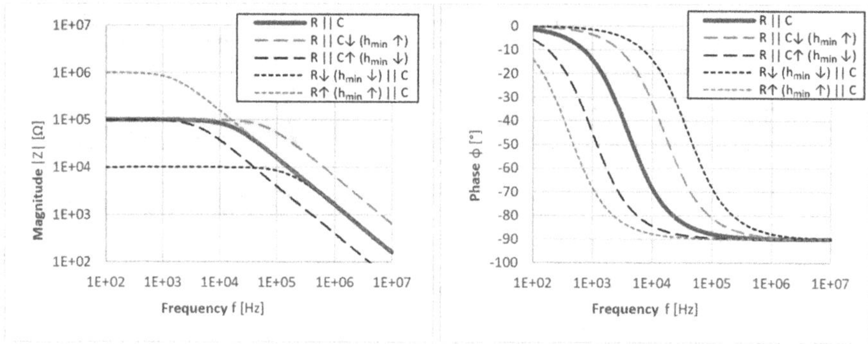

FIGURE 3.11 Typical magnitude and phase responses of a bearing impedance depending on the film thickness h_{min}.

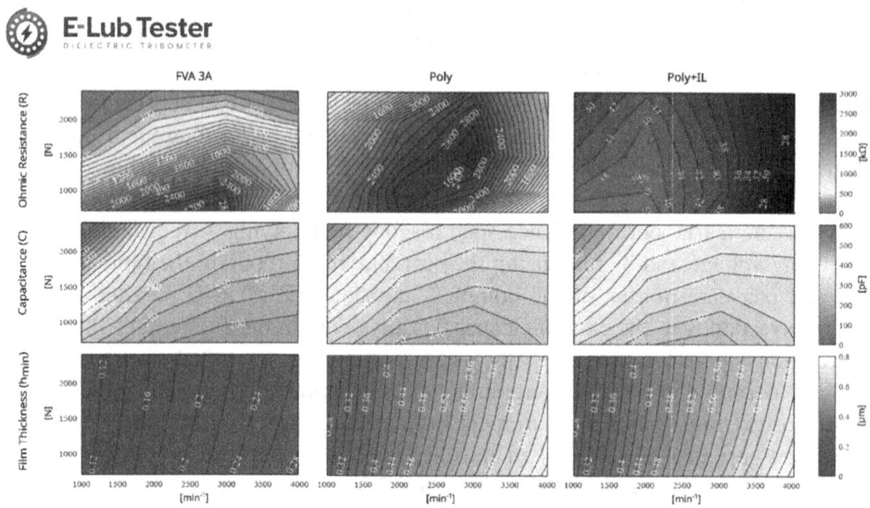

FIGURE 3.12 Bearing resistance, bearing capacitance and film thickness of FVA 3A mineral oil, polyglycol and conductive polyglycol (with 0.05 m% ionic liquid), depending on speed and axial load at 80°C.

The following measurements illustrate this difference. The three lubricant samples analyzed previously (Section 3.2.5) are subjected to characteristic map measurements with force and load variations in the dielectric tribometer. The values shown in Figure 3.12 are obtained. The addition of a small number of conductive molecules, in this case by adding 0.05 m% of ionic liquid, reduces the resistances (*R* values) of the polyglycol across the entire characteristic map to approximately 1% of the original values, whereas the *C* values remain practically unchanged. The lubricant film thickness (h_{min}) values given below were approximated according to Hamrock and Dowson's equation for point contact, based on measurements of the dynamic viscosity and the pressure viscosity coefficient in flucon's High-Pressure Fluid Laboratory [7].

It can be concluded that the bearing impedance is well suited to determine the condition of the bearing and lubricant and that it can be converted into the two parameters *R* and *C*, which are subject to different influences:

- **Ohmic Resistance *R***: The *R* value is strongly dependent on the condition of the lubricant. For example, if a pure base oil is initially a strong insulator delivering *R* values close to infinity, additives, contaminants, degradation products, or wear particles, etc., can lead to a measurable conductivity even in low concentrations, greatly reducing the *R* values, although rheologically, no influence on the lubricating properties can be observed (constant viscosity).

  In addition, a temporary drift of the *R* value can be observed as a result of electric discharges [8,9], which are investigated in the following Section 3.3.

  The pressure distribution in the lubricated bearing affects the measured overall ohmic resistance *R*, too. Since the electrical conductivity of a fluid typically decreases considerably through compression, it can be assumed that the effectively conductive lubricant quantity is distributed in a ring shape surrounding the EHL contact. Looking at Figure 3.10, this means that $R_{\text{Surrounding}} \ll R_{\text{EHL}}$.
- **Capacitance *C***: Based on the impedance measurements, a relationship between the capacitance and the lubricating film thickness can be established if the permittivity of the dielectric, i.e., the lubricant, is known. The capacitance indicates the lubrication condition of the roller bearing and is therefore the relevant parameter in many condition monitoring applications [10,11]. Without any further influences, a progression of the *C* value over the bearing service life, as shown in Figure 3.13, can be expected.

At the same time, the *C* value defines the electrical charge that would build up in the roller bearing if an external voltage existed in the application. This charge makes up a part of the amount of energy discharged in the event of an electrical breakdown (cf. Section 3.3.2).

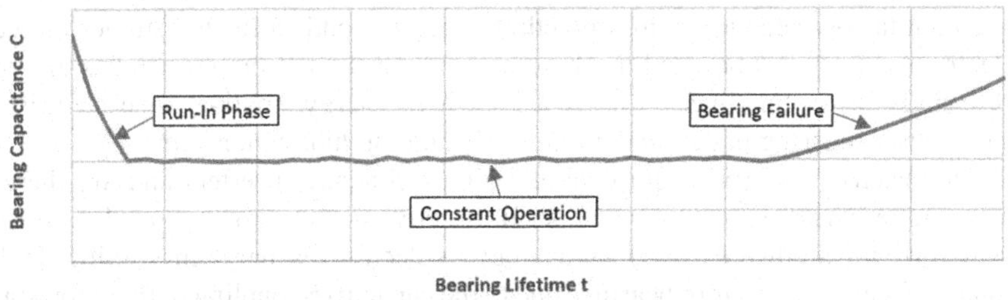

FIGURE 3.13 Condition monitoring based on $C = f$(bearing lifetime *t*).

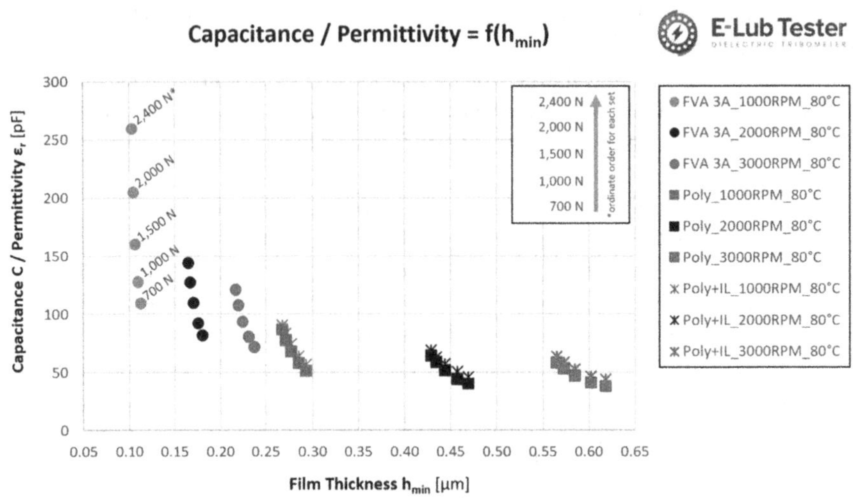

FIGURE 3.14 Permittivity-compensated capacitance vs. film thickness.

Lately, new attempts have been made to determine the film thickness directly through capacitance and permittivity measurements [12,13]. As can be seen in Figures 3.12 and 3.14, the capacitance strongly correlates with the film thickness. To exclude the influence of the dielectric fluid properties known from the pressurized capacitor tests, one can look at a permittivity-compensated capacitance value.

As shown above, impedance measurements and a separate analysis of the ohmic and the capacitive fluid behavior can be very helpful to analyze the lubrication conditions. Knowing the dielectric properties provided by the static capacitor tests, you will have a good starting point for looking into the dielectric strength of the lubricated electric drive.

3.3 DETERMINING THE DIELECTRIC STRENGTH

The *dielectric strength* is an indicator of the ability of a fluid to resist electrical stress. It gives information about the maximum electrical load a fluid can withstand before ionizing and having its electrical properties changed irreversibly. Whereas many traditional electrostatic investigations have focused on transformers, for which dissipation factors need to be minimized, the urge to conduct further studies on the dielectric strength of lubricated machine parts along the EV drivetrains has grown significantly in recent times. This is due to high-energy, spark-erosive electrical breakdowns occurring in the predominantly insulating lubrication gaps.

Many electric drivetrains are operated with frequency inverters and are therefore prone to harmful stray currents, such as eddy currents, rotor-ground currents and "EDM" breakdown currents (in analogy to electric discharge machining) [14]. Breakdown currents occur in bearings due to the capacitive coupling of the rotor, stator and motor windings and cause various damage patterns (see Figure 3.15).

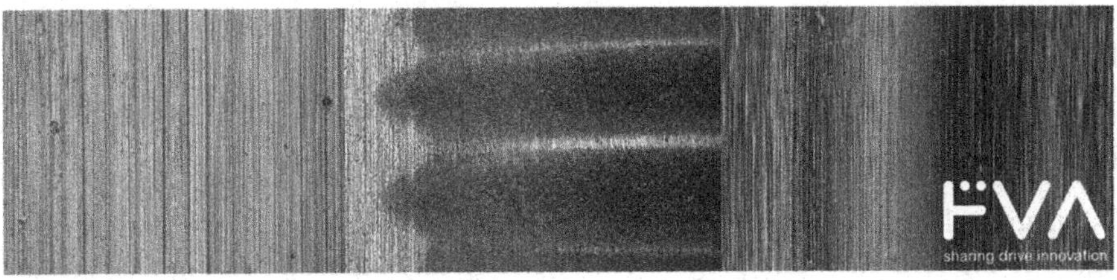

FIGURE 3.15 Raceway damage patterns from EDM breakdowns as investigated in the FVA 650 project [15] – *Left*: Craters/pitting, middle: ripples/fluting, *Right*: Greyfrosting.

3.3.1 Traditional Tests (Insulating Liquids)

Most traditional test methods regarding the dielectric strength are focused on the dielectric breakdown voltage. This is the maximum measurable voltage applied before breaking down. An overview over commonly used standards used to determine the dielectric breakdown voltage is given in Table 3.3.

As in the case of the dielectric properties (Section 3.2), these standards focus solely on electrically insulating fluids. Also, all these methods follow the same general principle: the insulating liquid is filled into a defined gap geometry between two electrodes. The electrode distance is usually in the range of a few millimeters. An AC voltage (typically in the range of several kiloVolts) is then applied to the electrodes and gradually increased until the voltage breaks down. The last measured voltage is then noted as the breakdown voltage.

Due to the statistical nature of the breakdown effect, it is easily understandable that the results of various measurements tend to scatter in ranges of almost 100%. Hence, a statistically relevant amount of measurements is required to give a comparable value between different samples.

While designed for insulating liquids like transformer oils, which have high breakdown voltages, these standards are ill-suited for further studies of moderately

TABLE 3.3
Commonly Used Standards for Determination of the Breakdown Voltage of Insulating Liquids

Designation	Title
ASTM D877	Standard test method for dielectric breakdown voltage of insulating liquids using disk electrodes
ASTM D1816	Standard test method for dielectric breakdown voltage of insulating liquids using VDE electrodes
ASTM D3300	Standard test method for dielectric breakdown voltage of insulating liquids under impulse conditions
IEC 60156	Insulating liquids – determination of the breakdown voltage at power frequency – test method

conductive EV fluids. Due to the vastly different dielectric properties of lubricants, differing breakdown voltages can be expected as well.

3.3.2 Newly Developed Lubricant Test Methods

As mentioned above, the dielectric strength is the decisive lubricant parameter with regard to the formation and progression of electrically induced bearing damages in the electric drivetrain. Nevertheless, no strict connection can be seen between the statically measured dielectric properties (electrical conductivity and relative permittivity) on the one hand and the dielectric strength on the other.

In principle, it should be noted that electrical breakdowns are sporadic events. This has already been considered in the test specifications of the traditional static tests and applies to an even greater extent to a dynamic application, e.g., when measuring a tribological contact: there are vibrations, periodic proximity effects of surface irregularities that trigger breakdowns resulting from the kinematics of the contact partners, as well as mixing and cavitation processes in the lubricant.

The main factor influencing the discharge tendency in the dynamic system is the thickness of the lubricant film and hence the minimum gap height between the electrodes of the electrified tribometer (see Figure 3.16).

Due to their sporadic occurrence, it is necessary to determine a statistically sufficient number of discharges to link them to possible influencing factors. The difficulty lies in the fact that the conditions for subsequent measurements are disturbed immediately after each individual event: as with all plasma discharge processes, an increased conductance temporarily occurs, which influences the voltage build-up at

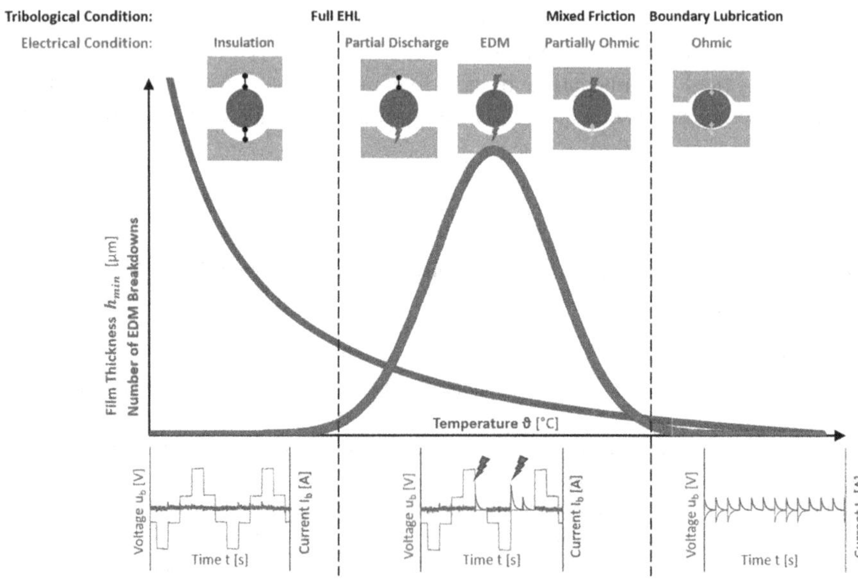

FIGURE 3.16 Discharge tendency of the tribo-contact [16,17].

the contact [8]. The metrological challenge is to generate high voltage amplitudes with very short rise times while simultaneously avoiding excessive current flow through the discharge channel. At this point, impedance spectroscopy also becomes a useful tool to assure the consistent condition of lubricated bearings immediately before a breakdown measurement.

Known factors influencing the dielectric strength are signal shape, vibrations, contained particles, contamination, and, of course, the lubricant with all its properties: e.g., viscosity, hygroscopicity, homogeneity, or the effectiveness of the additives in the gap [15,18].

At this point, it should be mentioned that the use of real roller bearings as test specimens offers two advantages. When measuring the dielectric strength, electrical breakdowns occur, leading to surface mutations of the rolling elements and – if the breakdowns occur frequently – to the falsification of subsequent electrical measurements [19]. Regular replacement of the test specimens is therefore highly recommended and can be achieved with cheap and mass-produced roller bearings. Furthermore, the shape of the electrodes greatly affects the dielectric strength [20]. Only in real roller bearings the typical pressure ellipses of ball-torus-shaped contact zones arise.

In the following sections, based on dielectric tribometer tests with the three lubricant samples already analyzed, measurement procedures for the dielectric strength of lubricants in roller bearings are described and at the same time the necessity of a separate consideration of R and C is explained.

3.3.2.1 Initial Breakdown Test (Non-Damaging Method)

In order to replicate the conditions of the electric drivetrain, the dielectric tribometer was equipped with a signal generator that can generate freely modulated voltages, such as a typical stair-shaped common-mode voltage, up to 60 V pkpk. By gradually increasing the voltage amplitude until the first breakdown, it is then possible to carefully approach the dielectric strength without significantly damaging the lubricant or the roller bearing. If a breakdown occurs, it can be identified by an offset of the periodic mean voltage even at a minimal current (Figure 3.17).

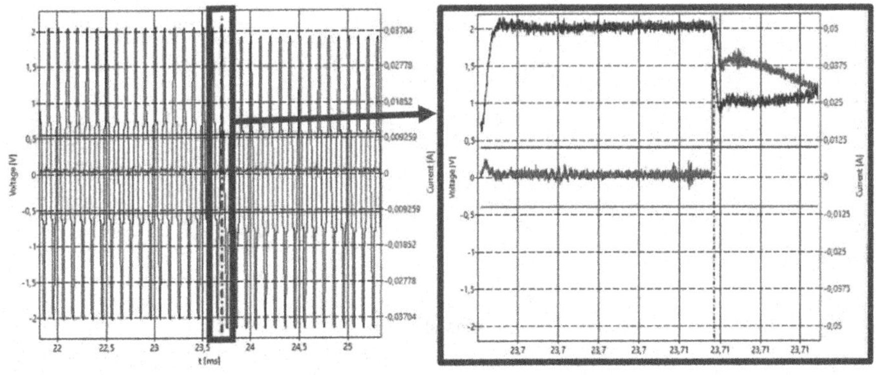

FIGURE 3.17 Voltage offset due to an initial breakdown.

Depending on the operating point, different film thicknesses are established, resulting in different breakdown distances. An automated, damage-free, patent-pending measuring method was developed [21] for multiple operating points, delivering coherent test results that correlate with the changes in the lubrication condition. With this method we try to master the various challenges of a precise determination of the dielectric strength.

3.3.2.1.1 *Challenge 1: Sporadic Occurrence of Breakdowns*

As mentioned above, discharges are sporadic events. The higher the voltage, the shorter the expected duration until a discharge occurs. If the observation period is limited by the measuring technology, it is not unlikely that discharges at low voltages will be overlooked. This problem can be countered, for example, by a debouncing algorithm. In an automated measurement process, a stop criterion is defined for further increase of the voltage amplitude, such as the repeated occurrence of discharge events in three consecutive measurements. This provides a voltage amplitude that is very likely to lead to breakdowns within the observation period (Figure 3.18, "Consistent Event"). At the same time, the "First Event" curve indicates that with a longer observation period, breakdowns at much lower voltages could also be regularly observed.

3.3.2.1.2 *Challenge 2: Consequences of Current Passage*

Immediately following a breakdown, there is a strong decrease of the resistance of the lubrication gap – the switch in the $R_{\text{Breakdown}}$ path is closed (cf. Figure 3.10). First, the breakdown generates an electric arc that forms plasma [22]. The resistance of the discharge channel can be measured and is orders of magnitude lower than the values of the remaining resistances $R_{\text{EHL}} || R_{\text{Surrounding}}$ [23]. Often, this effect not only persists until the voltage is reversed, but also causes another type of current flow: due to the remaining low resistance $R_{\text{Breakdown}}$, the voltage cannot be fully built up in the next

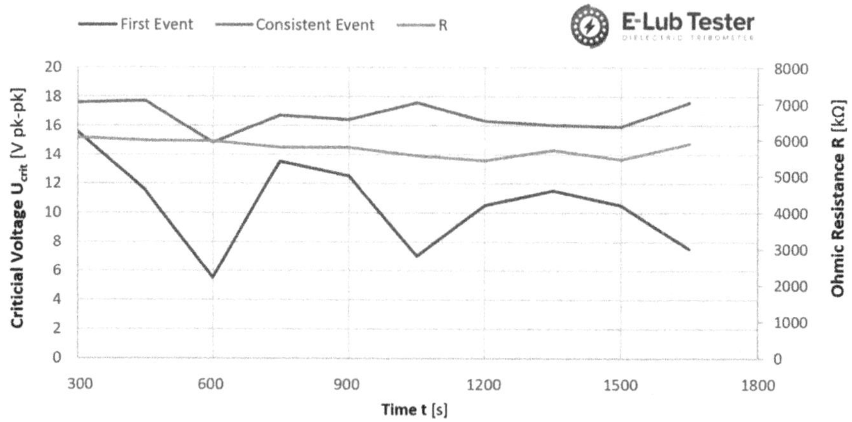

FIGURE 3.18 Consecutive determination of critical voltage amplitudes with and without a debouncing algorithm, with resistances measured beforehand.

signal period. An ohmic current flow occurs synchronously with a limited voltage build-up, maintaining the plasma channel.

Another common effect of an electrical breakdown is the formation of ionized molecules or of reaction products in the lubricant, which will decrease the resistance in the contact zone $R_{EHL} || R_{Surrounding}$ in the long term. It is assumed that recombination and mixing processes lead to a gradual return to the initial dielectric state, resulting in the impedance values originally measured. This recovery process depends heavily on the duration of the electrical exposure and on the operating conditions of the bearing [8].

Furthermore, strong discharges can cause permanent damage to the rolling element surfaces. This leads to local field line concentrations and thus to irreversible changes in the measured impedance values [19].

3.3.2.1.3 *Challenge 3: Influence of Vibrations and Surface Roughness*

The limited reproducibility due to manufacturing tolerances of both the test rig and the test bearings needs to be considered as well. The assembly must be qualified to prevent imbalances caused by the potentially cumulative eccentricities of the test bearing, shaft, center hole and drive shaft. Hence the dielectric tribometer is equipped with a vibration sensor to monitor the running smoothness.

Additionally, a position detection is established and synchronized with the breakdown measurements. This helps to localize surface peaks in the bearing raceways and potentially resulting proximity effects.

3.3.2.1.4 *How to Derive the Dielectric Strength from the Breakdown Tests?*

To exclude operating and geometric conditions, the breakdown voltage is related to the electrode distance. This leads to the dielectric strength E_{crit} for which the electrode distance corresponds to the approximated film thickness h_{min}:

$$E_{crit} = \frac{U_{crit}}{h_{min}} \tag{3.4}$$

It should be added that this is a simplified approach, also assuming a homogenous electric field. Gathering the critical breakdown voltages for the operating points investigated earlier (Section 3.2.6) gives us the results shown in Figure 3.19 for the three lubricants.

This allows us to calculate the critical field strength for the characteristic diagram to further investigate its speed and load dependency (see Figure 3.20).

3.3.2.2 Discharge Distribution Test (Damaging Method)

While the initial breakdown test described in the previous section uses a stop criterion to prevent discharge damages to the lubricant and bearing – and can therefore easily be added to a measuring program with multiple operating points – it is not a suitable method for determination of the discharge rate or to investigate the effects that repeated electrical breakdowns can have.

Therefore, a second method was developed for the *E-Lub Tester* dielectric tribometer that allows a potentially irreversible damage in order to analyze the

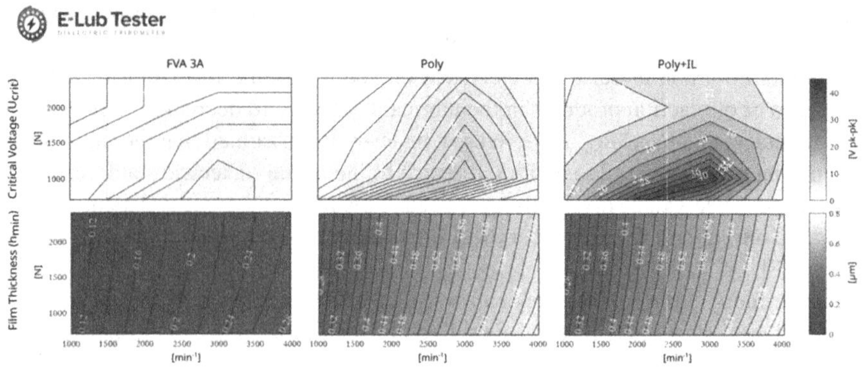

FIGURE 3.19 Breakdown voltage and film thickness of FVA 3A mineral oil, polyglycol and conductive polyglycol (with 0.05 m% ionic liquid), depending on speed and axial load at 80°C.

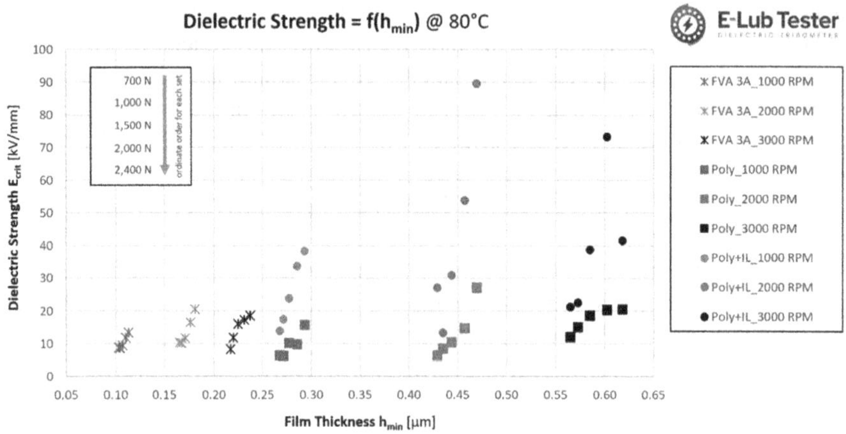

FIGURE 3.20 Dielectric strength vs. film thickness of FVA 3A mineral oil, polyglycol and conductive polyglycol (with 0.05 m% ionic liquid) at 80°C.

breakdown in more detail. The discharge distribution test is conducted at one suitable operating point, preferably providing fully elastohydrodynamic conditions. In this case we choose {80°C | 1000/min | 2.4 kN}. Then, a rectangle-shaped three-step AC source voltage is applied and gradually increased from 0.0 to 60.0 V pkpk in 0.5 V pkpk increments while measuring the voltage and the resulting current with a 125 MS/s sampling rate. Each voltage level is applied for 60 ms, equaling the duration of one full bearing rotation at 1000/min. The measured resulting currents can then be categorized into EDM breakdown currents and ohmic currents to determine their respective incidence and intensity distribution. Before proceeding to the next voltage level, zero voltage is applied for a recovery period of 180 seconds. In the case of full lubrication, the ohmic currents occur only after electrical discharges (cf. Section 3.3.2.1).

The graphs below show the discharge currents for each tested lubricant in different shades of grey that represent the amperage, starting with 100 mA, which in the case of the investigated 51208 thrust ball bearing was experimentally found to be the minimum intensity to cause pitting damage. Looking at the results of these three lubricants at an exemplary operating point, the respective discharge distribution curves (Figures 3.21–3.23) are found to have some remarkable similarities to one another: they all have in common that there is a maximum of breakdown currents at a certain source voltage. Further increasing the voltage will decrease the number of breakdowns and increase the amount of ohmic currents. An explanation for this would be that a conductive channel was formed through the discharge from the previous voltage level that remains active even after 180 seconds without further voltage applied.

Adding the ionic liquid will drastically lower the ohmic resistance of the polyglycol, and it also affects the breakdown tendency. In this case, initial 100 mA breakdowns occur at 6.5 V pkpk instead of at 3.5 V pkpk. This effect could be explained by the increased electrical conductivity that will dissipate some of the charges from the contact surfaces, preventing local hot spots in the electric field and increasing the overall dielectric strength.

Instead of using the same operating point, comparing FVA 3A mineral oil to the two polyglycols could be done under equidistant conditions through consideration of the difference in pressure viscosity. Using a new test bearing, we can look at another operating point {47°C | 1000/min | 2.4 kN} with an estimated equivalent film

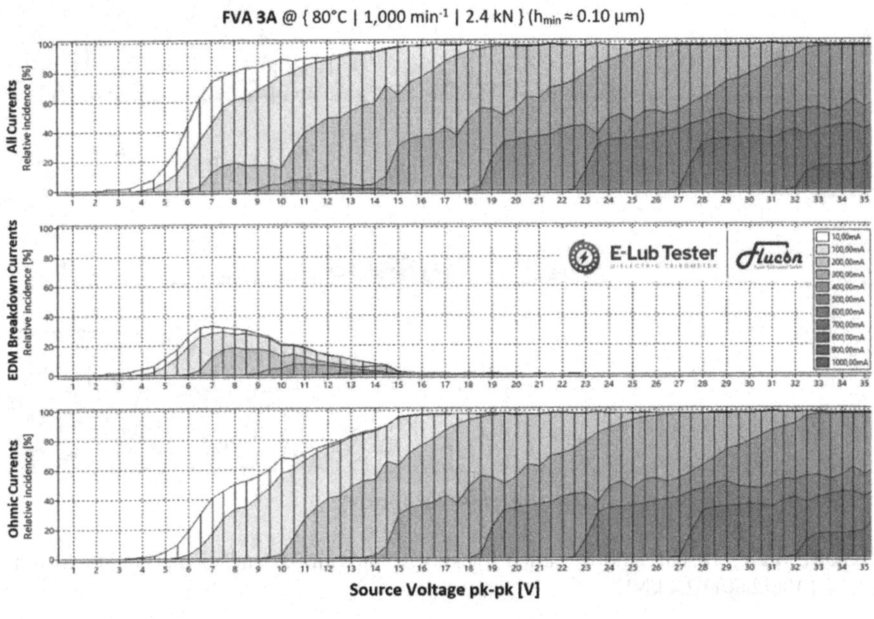

FIGURE 3.21 Discharge distribution for FVA 3A mineral oil at {80°C | 1000/min | 2.4 kN}.

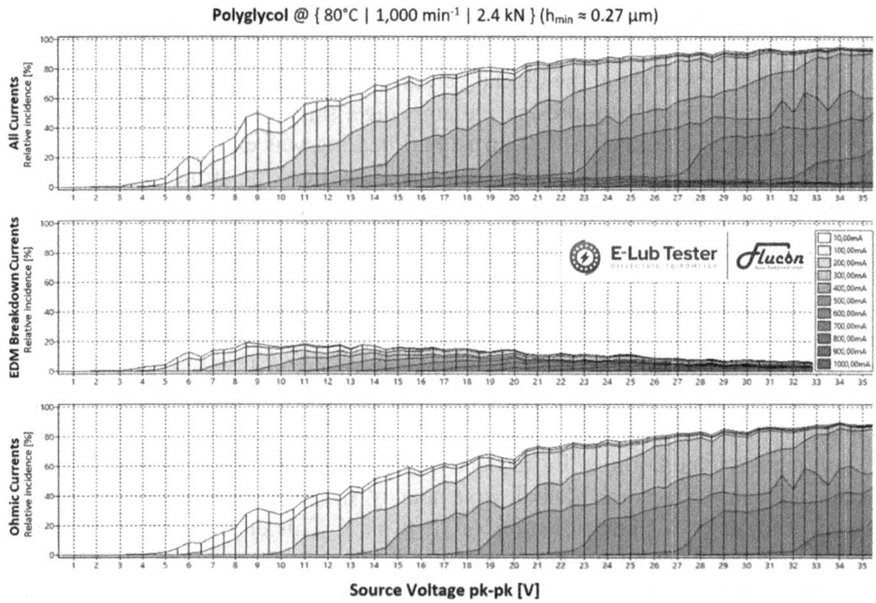

FIGURE 3.22 Discharge distribution for polyglycol at {80°C | 1000/min | 2.4 kN}.

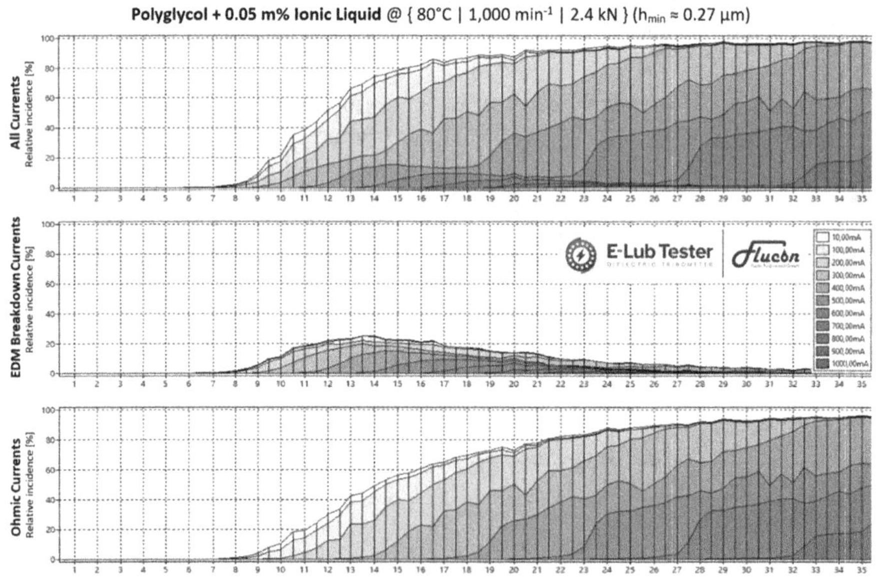

FIGURE 3.23 Discharge distribution for polyglycol with 0.05 m% ionic liquid added at {80°C | 1000/min | 2.4 kN}.

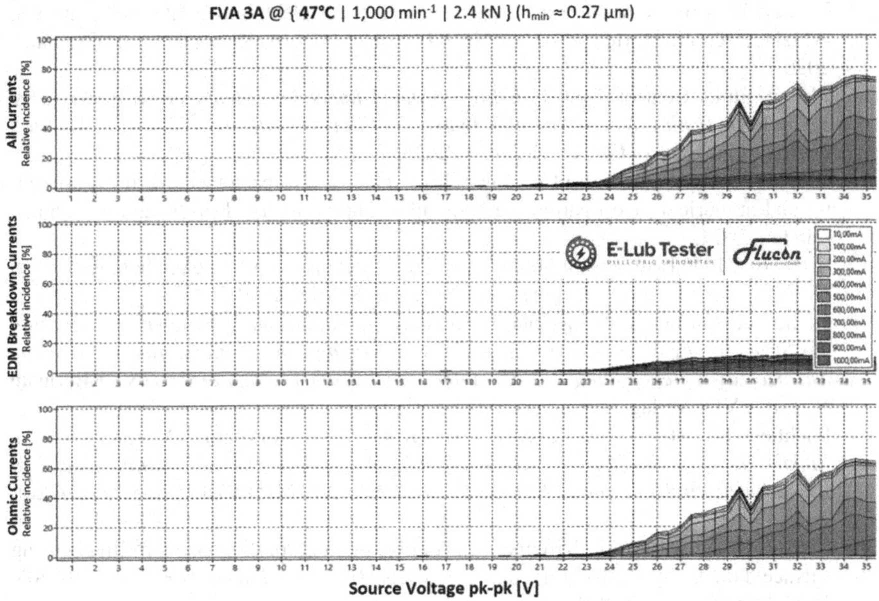

FIGURE 3.24 Results of equidistant test – discharge distribution for FVA 3A mineral oil at {47°C I 1000/min I 2.4 kN}.

thickness. The respective test results presented in Figure 3.24 show that FVA 3A now has a significantly higher dielectric strength than the polyglycols.

Studies such as these offer the potential to investigate the electrically induced damaging of EV fluids and mechanical components of the electric drivetrain. This will help to develop suitable countermeasures and to meet the challenges of tomorrow's drive technology.

REFERENCES

1. DIN 51111:2024-02, *Elektrische Eigenschaften von frischen und gebrauchten Ölen aus Elektroantrieben im Fahrzeug – Messung der spezifischen elektrischen Leitfähigkeit, der relativen Permittivität (εr) und des dielektrischen Verlustfaktors (tan δ)/Electrical properties of fresh and used oils from electric drives in vehicles – Measurement of the Specific Electrical Conductivity, the Relative Permittivity (εr) and the Dielectric Dissipation Factor (tan δ).* Beuth Verlag, Germany, 2024. https://doi.org/10.31030/3505323.
2. Flucon Fluid Control GmbH, *EPSILON+ Dielectricity Meter: Measurement of the Specific Electrical Resistivity and Conductivity, Relative Permittivity and Dielectric Dissipation Factor in Accordance with DIN 51 111 – Hardware & Software Manual.* Flucon Fluid Control GmbH, Barbis, 2024.
3. Hamrock, B. J., Dowson, D., *Ball Bearing Lubrication: The Elastohydrodynamics of Eliptical Contacts.* John Wiley and Sons, New York, 1981.

4. Bechev, D., Prüfmethodik zur Charakterisierung der elektrischen Eigenschaften von Wälzlagerschmierstoffen. PhD Thesis, TU Kaiserslautern, Kaiserslautern, Germany, 2020.
5. Flucon Fluid Control GmbH, *E-Lub Tester Dielectric Tribometer: Impedance & Breakdown Testing With A Variable Lubricant Film – Hardware & Software Manual.* Flucon Fluid Control GmbH, Germany, 2025.
6. Gonda, A., Determination of rolling bearing electrical capacitances with experimental and numerical investigation methods. PhD Thesis, RPTU Kaiserslautern Landau, Landau, 2024.
7. Bode, B., Modell zur Beschreibung des Fließverhaltens von Flüssigkeiten unter hohem Druck. *Tribologie und Schmierungstechnik*, 36, 182–189, 1989.
8. Heine, A., Hausner, S., Klingebiel, J., *Messung der komplexen Wälzlagerimpedanz und des Stromdurchgangverhaltens zur Charakterisierung von Schmierstoffen - Über neue Ansätze zur Untersuchung des variablen dielektrischen Schmierfilms*, VDI-Berichte Nr. 2415, VDI Verlag, Düsseldorf, 2023.
9. Puchtler, S., Maier, R., Kirchner, E., Kuhn, M., Burkhardt, Y., *Der Initiale Elektrische Durchschlag – Einflüsse und Auswirkungen/The Initial Electrical Breakdown – Influences and Effects,* VDI-Berichte Nr. 2415, VDI Verlag, Düsseldorf, 2023.
10. Martin, G., Becker, F. M., Kirchner, E., A novel method for diagnosing rolling bearing surface damage by electric impedance analysis. *Tribology International,* 170, 107503, 2022, https://doi.org/10.1016/j.triboint.2022.107503.
11. Becker-Dombrowsky, F. M., Schink, J., Frischmuth, J., Kirchner, E., Investigation of the electrical impedance signal behavior in rolling element bearings as a new approach for damage detection. *Machines*, 12, 487, 2024, https://doi.org/10.3390/machines12070487.
12. Flucon Fluid Control GmbH, Verfahren zur Ermittlung der Schmierfilmdicke zwischen zwei sich gegeneinander bewegenden Lagerpartnern sowie Vorrichtung zum Durchführen des Verfahrens. Flucon Fluid Control GmbH, Patent application 102024108 031.0, 2024.
13. Manjunath, M., Hausner, S., Heine, A., De Baets, P., Fauconnier, D., Electrical impedance spectroscopy for precise film thickness assessment in line contacts. *Lubricants,* 12(2), 51, 2024, https://doi.org/10.3390/lubricants12020051.
14. Muetze, A., Bearing currents in inverter-fed AC-motors. PhD Thesis, TU Darmstadt, Darmstadt, 2003.
15. Bechev, D., Weicker, M., Binder, A., *Schädlicher Stromdurchgang 2: Methodik zur praxisnahen Charakterisierung von elektrischen Schmierstoffeigenschaften zur Verbesserung der rechnerischen Vorhersage von Lagerströmen. Arbeitskreis Schmierstoffe und Tribologie, FVA-Forschungsvorhaben Nr. 650 II, Heft 1387.* Forschungsvereinigung Antriebstechnik e.V. (FVA), Frankfurt am Main, 2020.
16. Hausner, S., How shaft voltage causes bearing damage and lubricant degradation – a dual approach to electrical lubricant testing. In: *Presentation @ STLE Annual Meeting Minneapolis*, Minneapolis Convention Center in Minneapolis, Minnesota, 2024.
17. Radnai, B., Wirkmechanismen bei spannungsbeaufschlagten Wälzlagern. Maschinenelemente und Getriebetechnik Berichte Bd. PhD Thesis, TU Kaiserslautern, Kaiserslautern, Germany, 2016.
18. Arbeitskreis Schmierstoffe und Tribologie, *Stromdurchgang in Wälzlagern: Untersuchung von Oberflächenveränderungen und Folgeschäden an Wälzoberflächen durch Stromdurchgang*, FVA-Forschungsvorhaben Nr. 650 III, Heft 1614, Forschungsvereinigung Antriebstechnik e.V. (FVA), Germany, 2024.
19. Klingebiel, J., Hausner, S., Schröder, M., Change in impedance as a result of surface mutations induced by parasitic current transfer. In: *Bearing World Conference Report*, Würzburg, Germany, 2022.

20. Schumann, W. O., *Elektrische Durchbruchfeldstärke von Gasen: Theoretische Grundlagen und Anwendung*. Springer, Berlin, Heidelberg, 1923. https://doi.org/10.1007/978-3-662-25891-0.
21. Flucon Fluid Control GmbH, Verfahren zur Ermittlung einer Durchschlagsspannung eines Schmierstoffes in einem bewegten Lager sowie Vorrichtung zur Durchführung des Verfahrens. Flucon Fluid Control GmbH, Patent application 102024108035.3, 2024.
22. Gemeinder, Y., Lagerimpedanz und Lagerschädigung bei Stromdurchgang in umrichtergespeisten elektrischen Maschinen. PhD Thesis, TU Darmstadt, Darmstadt, 2016.
23. Graf, S., Charakterisierung und Auswirkungen von parasitären Lagerströmen in Mischreibung. PhD Thesis, RPTU Kaiserslautern, Landau, 2023. https://doi.org/10.26204/KLUEDO/7407.

4 Degradation Testing
Polymer Compatibility

Yungwan Kwak

ABBREVIATIONS

ACEA	European Automobile Manufacturers' Association
ANSI	American National Standards Institute
ASTM	American Society for Testing and Materials
ATF	Automatic Transmission Fluid
BDV	Breakdown Voltage
CEC	Coordinating European Council
DMTA	Dynamic Mechanical Thermal Analysis
EDX	Energy Dispersive X-ray Spectroscopy
EPDM	Ethylene-Propylene-Diene Monomer
ETF	Electric Transmission Fluid
EV	Electric Vehicle
FKM	Fluoroelastomer
FTIR	Fourier Transform Infrared Spectroscopy
HEV	Hybrid Electric Vehicle
IEC	International Electrotechnical Commission
ISO	International Organization for Standardization
MDPI	Multidisciplinary Digital Publishing Institute
MW	Magnet Wire
NEMA	National Electrical Manufacturers Association
OEM	Original Equipment Manufacturer
PA6	Polyamide 6
PA66	Polyamide 66
PAI	Polyamideimide
PAO	Polyalphaolefin
PBI	Polybenzimidazole
PBO	Polybenzoxazole
PDIV	Partial Discharge Inception Voltage
PEEK	Polyether Ether Ketone
PEI	Polyetherimide
PI	Polyimide
POE	Polyol Ester
PPQ	Polyphenylquinoxaline
PTFE	Polytetrafluoroethylene
RPDIV	Repetitive Partial Discharge Inception Voltage

DOI: 10.1201/9781003496069-4

SAE	Society of Automotive Engineers
SEM	Scanning Electron Microscopy
TGA	Thermogravimetric Analysis
XRD	X-ray Diffraction

4.1 INTRODUCTION

Electric vehicles (EVs) and hybrid electric vehicles (HEVs) are becoming increasingly prevalent in today's automotive landscape. This shift is driven by government regulations and consumer demand, with a growing emphasis on sustainability, energy efficiency, and pollution reduction. These vehicles rely on advanced non-metallic materials, such as engineering plastics, seals, elastomers, electrical insulation materials, and lubricants, to meet the demands of electrified drivetrains and ensure optimal performance, efficiency, and durability.[1,2,3] It is worth mentioning that the terminology for non-metallic materials, polymers, and plastics was used interchangeably in this chapter with a broader meaning.

Engineering plastics, mostly used as structural materials, play a crucial role in automotive applications by providing lightweight, durable, and corrosion-resistant components. These plastics are found in gears and bearing parts, high-voltage systems, wiring harnesses, connectors, component casings, etc. Seals and elastomers are also crucial, as they keep lubricants and fluids contained within designated areas. Compromised integrity of seals and elastomers can lead to fluid leaks that may cause serious issues, sometimes even catastrophic vehicle failures. Insulation materials are indispensable for controlling and enhancing the efficiency and longevity of vehicles by providing proper insulation to conductors, typically including magnet wires (MWs), as well as low and high voltage connectors and cables.

Compatibility issues between the aforementioned non-metal materials and lubricants can significantly affect hardware integrity and vehicle performance.[4,5,6] Achieving optimal lubricant performance in HEVs and EVs requires careful consideration of polymer compatibility with lubricant additives. Hardware designers must address the compatibility of non-metallic materials that not only directly encounter lubricants but also come into contact with vapors or mists under various temperature cycles.

This chapter provides a concise review of compatibility tests for various non-metallic materials with electric transmission fluids (ETFs) designed for HEV and EV applications. The author will also discuss examples of material compatibility tests and provide insights into the future directions of research.

4.2 POLYMERS IN HEV AND EV DRIVELINE SYSTEMS

Figure 4.1 includes the composition of typical electric motors used for HEV and EV applications. The left-hand side of Figure 4.1 shows a round wire motor, while the right-hand side of Figure 4.1 depicts a hairpin electric motor. As shown, these electric motors incorporate various non-metal components, such as insulation paper, varnish resin, tie cord, MW coating, electrical wires, resolvers, and low voltage connectors; however, the latter two are not depicted in the Figure 4.1.

FIGURE 4.1 (*Left*) An image of a random-wound motor and (*right*) an image of hairpin-wound motor. (Provided courtesy of Afton Chemical Corporation.)

Insulation materials play a critical role in ensuring safety and efficiency, as they come into direct contact with lubricants and are subjected to a wide range of operating temperatures. Degradation of MW, a polymeric insulation-coated copper wire, can result in current leakage, accelerated wear of bearings and gears due to electrostatics, grounding issues, and short-circuiting.[5,6] These issues may ultimately lead to failures in both the electric motor and the vehicle. The resolver, which detects motor speed and the position of magnets, communicates with the electronic control unit to regulate vehicle speed. A failure in this component can cause motor response problems, efficiency loss, or even catastrophic failure. Low voltage connectors, which link the resolver, power supply, and sensors, must be compatible with lubricants to ensure proper signal transmission to the resolver, thereby preventing wire corrosion, degradation, and conductive deposit formation on the connector pins or MW.

Additional electrical insulation and mechanical strength are provided by insulation paper, binding rope, and varnish resin. Insulation paper is crucial for ensuring that round or rectangular MWs are inserted to slots correctly without scratches, and it prevents the MWs from contacting the stator core's metal laminations, which could lead to short-circuiting and motor failure. Varnish resin is applied to secure the MW assembly in place and to protect it from mechanical or chemical damage due to vibration, lubricants, or other contaminants, while also enhancing the insulation properties of the MWs.[6]

4.3 COMPATIBILITY CHALLENGES

As previously mentioned, various types of plastics encounter lubricants through direct liquid, mist, or vapor contact in driveline of HEV and EV. These lubricants include several additives such as dispersants, detergents, friction modifiers, antioxidants, seal swell agents, antifoaming agents, extreme pressure agents, and anti-wear agents, etc. in addition to base oils. The heat profiles these plastics experience can vary significantly, with materials like stator coil insulation facing temperatures up to

180°C or higher. A notable example is the MW in the stator coils of electric motors. This wire undergoes extreme thermal conditions that are exacerbated by the operational stress of the motor. Moreover, these plastics are exposed not only to thermal stress but also to significant vibrations during operation. Consequently, it is essential for the plastic materials to demonstrate robust chemical and mechanical compatibility with the lubricants, ensuring reliable performance under the harsh conditions of vehicle operation over the vehicle's entire lifespan.

Seals and elastomers within these systems are also subjected to rigorous high and low temperature cycles and vibrations. Maintaining their integrity is critical; a loss in cross-linking density or the formation of cracks can render them brittle, thereby compromising their mechanical strength and leading to leaking of fluids and potential failures in sealing performance. Ensuring compatibility and resilience of these materials is crucial for maintaining the overall reliability and efficiency of the drive unit and the vehicle.

As the automotive industry evolves, new lubricants specifically designed for HEVs and EVs are being developed. However, many HEVs and EVs currently utilize automatic transmission fluid (ATF) style lubricants composed of various components. While some of these components are benign and do not affect the integrity of plastics, others can be detrimental. They can degrade plastics by reducing cross-linking density, leaching out plasticizers, swelling, or breaking chemical bonds, which compromises the materials' structural integrity and functionality.

4.4 STANDARD TEST PROCEDURES FOR NON-METALLIC MATERIAL COMPATIBILITY TESTS

While additional standard test methods may exist, the following procedures cover the most utilized tests for non-metallic materials discussed throughout this chapter. These tests are divided into four categories: plastics, insulating materials, rubber and elastomers, and MWs. Conducting these tests is crucial for determining the suitability of these materials for intended applications, ensuring that they meet the required standards of performance and safety. SAE J3200,[7] published by the Society of Automotive Engineers (SAE) and developed by the TC 3 committee, provides guidelines for conducting plastics compatibility tests among other eMobility fluid requirements. Assessing compatibility between plastic materials and lubricants is one of the sections described in detail. It addresses various engineering plastics, including polyamides, polyphenylene sulfide, and polybutylene terephthalate, as well as adhesives, insulating paper, and polymer fabrics, etc. The document includes few of relevant tests detailed below. It outlines procedures for preparing test specimens, exposing them to lubricants under controlled conditions, and evaluating effects such as changes in appearance, dimensions, and mechanical properties.

4.4.1 Plastics

ASTM D638[8] – Standard test method for tensile properties of plastics: This test method defines test specimens and measures the tensile properties of plastics, which include tensile strength, elongation, yield point, and modulus of elasticity. The tensile strength is calculated as the maximum stress the specimen can withstand, while elongation at break measures the amount of deformation before failure. The modulus of elasticity indicates the material's stiffness in the elastic deformation range. It's used to determine how a plastic material will react when it is pulled apart or stretched, providing crucial information on its suitability for use in different applications.

ASTM D882[9] – Standard test method for tensile properties of thin plastic sheeting: This method specifically addresses the tensile testing of thin plastic sheeting, typically less than 1 mm in thickness. This test procedure describes various techniques for specimen gripping, extension measurement, grip separation, and test speed. Typical plastic properties such as tensile stress, elongation and modulus are measured.

ISO 175[10] – Plastics – Methods of test for the determination of the effects of immersion in liquid chemicals: This test assesses how plastic materials withstand to chemicals by immersing them in specified chemicals for a set period. It evaluates changes in physical properties like mass, dimension, strength, and appearance, which is vital for applications where chemical exposure is common.

ISO 527[11] – Plastics – Determination of tensile properties: This standard is similar to ASTM D638. This test method is used to determine the tensile properties of plastics such as tensile strength, elongation at break, and modulus of elasticity. The test involves preparing standardized dumbbell-shaped specimens, which are then stretched at a constant rate until they break. ISO 527 provides standardized procedures for comparing and evaluating the mechanical properties of various types of plastic materials, including rigid, semi-rigid, and flexible materials.

4.4.2 Rubbers and Elastomers

These tests are fundamental for assessing the durability and functionality of rubber and elastomer products in their respective environments, ensuring that they meet the required standards for performance and safety.

ASTM D412[12] – Standard test methods for vulcanized rubber and thermoplastic elastomers – tension; this test evaluates the tensile (tension) properties of vulcanized thermoset rubbers and thermoplastic elastomers. Specimens tested may be in the shape of dumbbells, rings or straight pieces of uniform cross-sectional area. It involves stretching an elastomer at a constant rate until it breaks, recording parameters like tensile strength, elongation at break, and modulus.

ASTM D471[13] – Standard test method for rubber property – effect of liquids: This test method evaluates the effect of liquids on rubber by measuring changes in mass, volume, and tensile strength after immersion in a controlled liquid environment. It is commonly used to determine how different fluids, such as oils or solvents, affect rubber materials, which is critical for applications where rubber parts are exposed to various chemicals.

ASTM D2240[14] – Standard test method for rubber property – durometer hardness: This standard describes the procedure for measuring the hardness of rubber, elastomers, or similar materials using a durometer. Hardness, in this context, is defined as the resistance to indentation under standardized conditions. This test is essential for determining the suitability of rubber for specific applications where hardness impacts performance, such as in seals or gaskets.

ASTM D6515[15] – Standard test method for rubber shaft seals determination of recovery from bending: This standard test method determines the recovery response of rubber after specific bending deformation. It assesses the relative performance potential of rubber compounds used in the manufacture of shaft seals. The test involves aging the bent rubber, whose ends are clamped, in fluids at a specified temperature and time. Valuable insights are gained for material selection and seal fabrication by measuring the distance between the ends of each specimen.

ASTM D7216[16] – Standard test method for determining automotive engine oil compatibility with typical seal elastomers: This test method assesses the compatibility of engine oils with automotive seal materials, specifically focusing on changes in the properties of elastomers, such as volume change, durometer hardness, and tensile strength. It helps in determining whether a particular engine oil might cause swelling, shrinking, or degradation of elastomeric seals, which is crucial for maintaining the reliability and longevity of automotive engines.

ISO 1817[17] – Rubber, vulcanized or thermoplastic – determination of the effect of liquids: Similar to ASTM D471, ISO 1817 tests the resistance of vulcanized rubber to various liquids by observing changes in properties like volume, mass, and hardness after immersion. It's widely used in industries where rubber materials are expected to perform under chemical exposure including automotive industry.

CEC L-112-16[18] – ACEA seals compatibility: This test, specified by the European Automobile Manufacturers' Association (ACEA), evaluates the compatibility of lubricants with elastomer materials used in automotive seals. It involves exposing materials to lubricants under specified conditions to see how well they resist physical and chemical changes. This test is essential for the automotive industry to ensure that lubricants do not compromise the integrity of engine and transmission seals.

4.4.3 Insulating Materials

These standards provide critical guidelines for testing the physical and electrical properties of various insulating materials, ensuring they meet safety, performance, and quality specifications for their intended uses.

ASTM D149[19] – Dielectric breakdown voltage and dielectric strength of solid electrical insulating materials at commercial power frequencies: This standard test method measures the dielectric breakdown voltage (BDV) and strength of solid electrical insulating materials under specified conditions. The test involves applying a voltage at a controlled rate to a sample until a breakdown occurs. The results help assess the material's ability to withstand electrical stress without failure.

ASTM D2519[20] – Bond strength of electrical insulating varnishes by the helical coil test: This test measures the bond strength of electrical insulating varnishes used on MW and other similar applications. It typically involves preparing a set of test specimens, which is helical coil of aluminium or copper, applying the varnish, and then testing the strength of the bond under specified conditions to ensure the varnish can adequately secure the wire insulation.

IEC 60819-2[21] – Non-cellulosic papers for electrical purpose: This standard is part of a series that specifies methods for testing non-cellulosic papers for electrical purposes. IEC 60819-2 specifically covers methods for evaluating properties like thickness, tensile strength, elongation at break, and moisture content among others. It's essential for determining the suitability of insulating paper used in electrical insulation systems.

ISO 1924-2:2008[22] – Paper and board – Determination of tensile properties, constant rate of elongation method (20 mm/min): This test describes methods for measuring the tensile strength, elongation at break, and other tensile properties of paper and board. This standard is crucial for evaluating the mechanical strength of these materials, which is important in packaging, construction, and other applications requiring paper and board materials.

4.4.4 Magnet Wires

ANSI/NEMA MW 1000[23] – American national standard for MW: NEMA MW 1000 is a comprehensive standard set by the National Electrical Manufacturers Association (NEMA) for MW, covering specifications, testing procedures, and requirements for both copper and aluminium wire in round and rectangular shapes. It addresses various test methods for key characteristics such as insulation quality, thermal endurance, and mechanical properties to ensure that MW performs reliably in applications, such as in motors and transformers. This standard is essential to ensure consistent quality and performance in electrical and electronic devices.

ASTM D2307[24] – Thermal endurance of film-insulated round MW: This standard test method evaluates the thermal endurance of film-insulated round MW. It involves cyclically heating the wire and testing MW's electrical property at intervals until it fails to meet specified criteria. The results help in determining the relative lifespan of the wire under continuous operation at various temperatures in air at atmospheric pressure. This test method is similar to IEC 60172 but differences exist regarding test specimen preparation.

ASTM D1676[25] – Standard Test Methods for Film-Insulated MW: It broadly covers the methods for testing film-insulated MWs, including round, rectangular or square, copper or aluminium, which are used primarily in electric motors, transformers, and other electromagnetic devices. The standard includes tests for evaluating properties such as bonding strength, continuity of insulation, adherence and flexibility of insulation, and thermal properties. Many test methods described here can be also found on NEMA MW 1000. This comprehensive testing ensures the wire meets specific performance and reliability requirements.

IEC 60172[26] – Test procedure for the determination of the temperature index of enamelled and tape wrapped winding wires: This standard outlines the procedure for assessing the thermal endurance of enamelled and winding wires. It involves subjecting the wire to elevated temperatures at atmospheric pressure and observing the time it takes for the insulation properties to degrade to predefined failure criteria. The temperature index is a value that represents the maximum operating temperature at which the wire can function for a standard period without significant loss of its insulating properties. This information is valuable for designers and development engineers when selecting the appropriate MW for electrical equipment and similar applications. This test method evaluates both round MWs as well as rectangular or hairpin MWs.

4.5 POLYMER COMPATIBILITY AND DEGRADATION TESTS

4.5.1 Plastics Test

Tormos et al.[27] applied both ASTM D7216 for the test procedure and ASTM D638 for the test parts preparation. They studied the effects of base oils on polyamide 6 (PA6) through immersion tests in polyalphaolefin (PAO) and polyol ester (POE). The results revealed significant impacts on PA6's physical and chemical properties, including mass and hardness changes, tensile strength and elongation changes, color change, FTIR, and crystallinity by XRD. Interestingly, POE showed milder influence compared to the other fluid, indicating their potential as superior base oils for thermal management in electric vehicle batteries. However, when selecting thermal fluids for EV applications, it's crucial to consider all other critical properties beyond material compatibility, including electrical properties, antioxidation performance, corrosion protection, etc. Therefore, choosing the right lubricant additive system in addition to the base oil is critical for developing suitable fluids for complex practical applications.

Garcia Tuero et al.[28] investigated the compatibility of structural polymers with four conventional ATFs using the ASTM D7216 soaking method at 100°C for up to 672 hours. Some of the results here were also discussed in later publication.[29] Figure 4.2 below illustrates the performance of polyether ether ketone (PEEK) at various aging times as an example of the results.

PEEK and polytetrafluoroethylene (PTFE) demonstrated full compatibility in both short and long-term tests regarding volume, hardness, and tensile strength properties with all ATFs used. However, polyamide 66 (PA66) exhibited poorer compatibility with the oils compared to the other materials, decreasing tensile and elongation strength. This was attributed to an increase in crystallinity, as measured by XRD, which altered the mechanical properties of the aged PA66 samples. FTIR results indicated that variations in the mechanical properties of the materials were generally more influenced by aging time than by the type of oil, except for PTFE, which was relatively unaffected by aging time. Overall, the ATFs showed good compatibility with the studied polymers, with PA66 being the only material to show a significant decline in elongation at break after aging.

The study emphasized that although these results indicate compatibility, further comprehensive tests are necessary to confirm their suitability for use in electrified transmissions. Additional oil compatibility tests, including electrical, magnetic, thermal, corrosion, aeration, and foaming tests at high speeds, must be performed. It is also important to note that the results of these tests should not be used to predict the lifetime of a component in real-world applications.

ISO 175 test procedure was also employed to study the compatibility of commonly used materials in EVs and HEVs. Higuchi et al.[30] investigated two types of polyamides and one type polyphenylene sulfide by soaking them in three different HEV

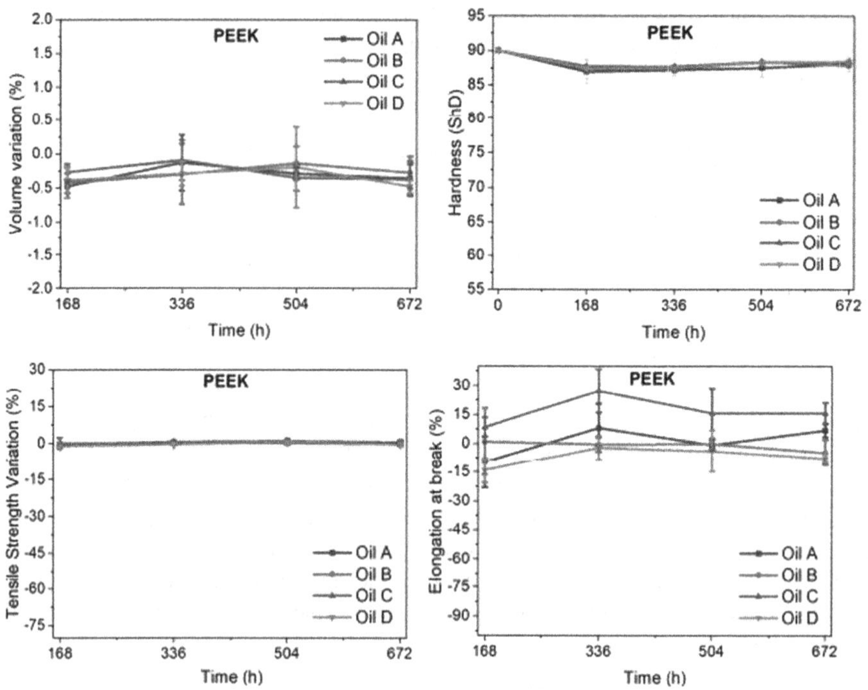

FIGURE 4.2 The changes in volume, hardness, tensile strength, and elongation of PEEK after immersion in oils, A–D at 100°C.[28] (Reproduced with guidelines from MDPI's open access policy. Copyright 2022, MDPI.)

fluids at 150°C for 1008 hours. They found that the compatibility of these materials varied depending on both the type of material and the type of lubricant, as indicated by changes in tensile strength at break. This testing has proven useful in comparing the responses of different plastic types to HEV fluids under similar conditions.

4.5.2 Rubbers and Elastomers Test

The static seal compatibility test is typically conducted under fluid immersion conditions using dog bone-shaped specimens to measure changes in volume, hardness, elongation at break, and tensile strength. It is important to note that static seal compatibility does not necessarily correlate well with dynamic seal compatibility due to different operating conditions. Static tests typically do not account for factors such as aeration and oxidation effects, which are more prevalent in dynamic seal applications.[31]

Bennett et al.[32] evaluated the impact of chemical attack by various amines on fluoroelastomers through tensile testing over time, dynamic mechanical thermal analysis (DMTA), and elemental analysis using energy dispersive X-ray spectroscopy (EDX). The kinetics of attack for each amine correlated with changes in cross-link density and diffusion effects. Notably, chain scission was found to play a more significant role

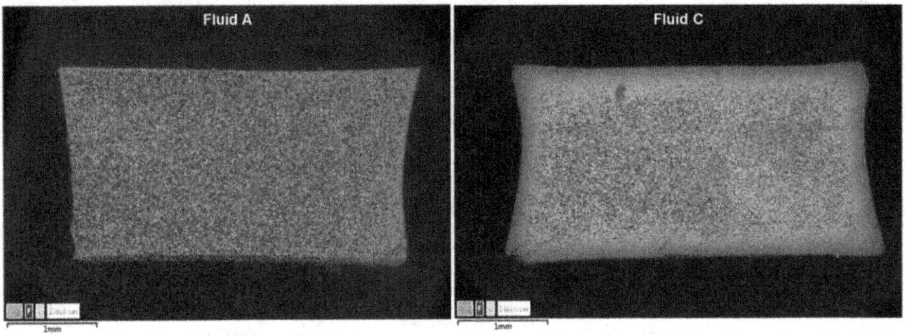

FIGURE 4.3 EDX image of the cross-section of a fluoroelastomer dogbone exposed to fluid A (*left*) and to fluid C (*right*) for 14 days at 150°C.[32] (Used with permission of SAE International, from[32]; permission conveyed through Copyright Clearance Center, Inc.)

in diminishing the physical properties of the elastomers than previously recognized. Furthermore, elastomers that underwent increased cross-linking exhibited enhanced elasticity, maintaining more of the bulk polymer network as demonstrated by tensile test measurements, compared to those affected by chain scission. Figure 4.3 below illustrates a technique for quantifying the fluorine content in fluoroelastomers using elemental mapping via EDX.[29] Fluid A, which performed well, maintained an overall cross-section appearance similar to that of the new elastomer, with consistent fluorine distribution. In contrast, poorly performing Fluid C exhibited noticeable fluorine depletion on the cross-section's surface, providing direct evidence of the dehydrofluorination mechanism. The authors observed that fluoroelastomer compatibility improved with larger molecular weight dispersant components, whereas shorter molecular weight amines showed greater aggressiveness in degrading the fluoroelastomer.

Garcia et al.[29,33] investigated the compatibility of elastomers with ATFs, evaluating three common types: fluoroelastomer (FKM), ethylene-propylene-diene monomer (EPDM), and vinyl-methyl silicone rubber (silicone). The aging process followed ASTM D7216-15 standard procedures, immersing the materials at 100°C for up to 667 hours.

Figure 4.4 illustrates specific test results for silicone elastomer exposed to four different types of ATFs. Notably, certain fluids demonstrate greater compatibility with silicone elastomer than others. These findings underscore the influence of ATF compositions on elastomer properties, such as swelling, hardness, and other mechanical characteristics. This understanding is valuable for lubricant researchers aiming to design ATFs that maintain compatibility with elastomer materials, thereby ensuring efficient and durable transmission performance.

Both dynamic and static seal compatibility are crucial in their respective applications. Dynamic seals undergo tests involving load, rotation, and varying temperatures, whereas static seals tests described above are exposed to high temperatures while soaked in fluid. Figure 4.5[34] below illustrates an example test apparatus for dynamic seals subjected to the conditions for typical dynamic seals' test.[31] To meet the demands of new hardware complexity and severe operating conditions, dynamic

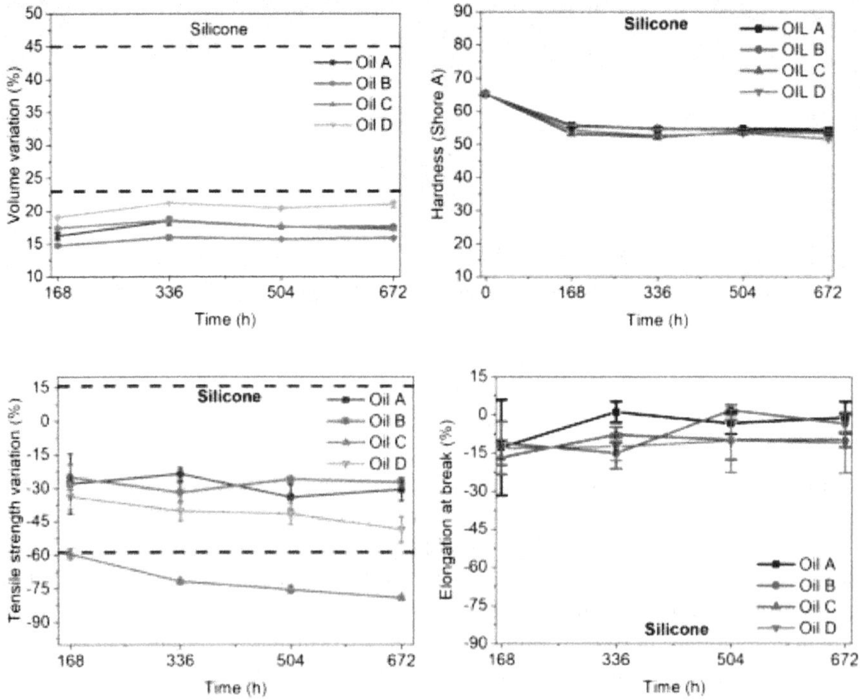

FIGURE 4.4 Silicone compatibility results (% volume change, hardness change, % tensile strength variation, and % elongation change at break) in four different lubricants after soaking up to 672 hours at 100°C.[33] Reproduced with guidelines from MDPI's open access policy. (Copyright 2022, MDPI.)

FIGURE 4.5 FKM dynamic seal test apparatus.[34] (Provided courtesy of Afton Chemical Corporation.)

seal tests were conducted for up to 2016 hours at 70°C in this case, with the shaft rotating in both directions and the shaft speed continually varying. Under these conditions, lubricant fluids must dissipate heat to reduce temperatures and lower friction, provide effective lubrication, and be compatible with dynamic seals.

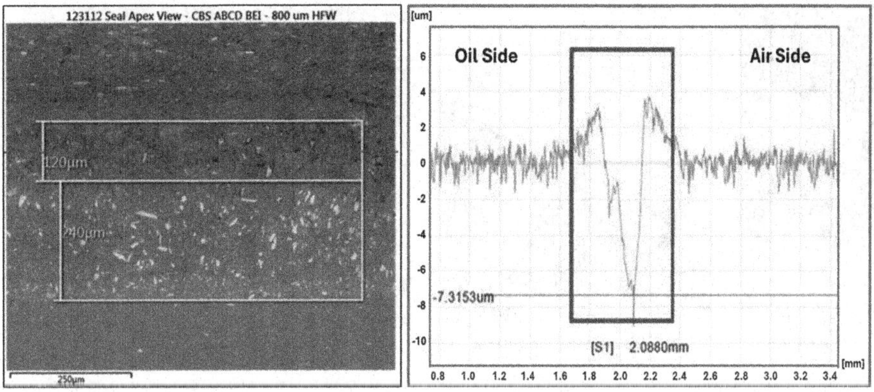

FIGURE 4.6 SEM/EDX analysis of the dynamic seal (left) and the surface profile analysis for the shaft (right).[35] (Provided courtesy of Afton Chemical Corporation.)

SEM-EDX analysis serves as an effective tool for investigating deeper degradation phenomena, as illustrated in Figure 4.6 above. This method complements typical mechanical tests, such as hardness, tensile strength, and elongation tests. These analyses reveal that deeper scars expose silicate filler particles, which contribute to wear. Additionally, areas with greater wear exhibit either thin or absent tribofilms. The shaft surface exhibited noticeable wear, as depicted by profilometry analysis shown in the figure on the right-hand side below.

4.5.3 Magnet Wires and Insulation Materials Tests

Typically test procedures are somewhat different for round MW and hairpin (rectangular or square) MW, although they are testing very similar properties required as MWs.

Kim et al.[35] evaluated polyetherimide (PEI) and polyamideimide (PAI) for MW application to consider their flexibility on Cu sheet. Their in-house test concluded that PEI showed better performance compared to PAI varnish. By performing a folding test, a lot of cracks were observed for the PAI coating, which means that the interface between the PAI and Cu sheet was damaged. PEI varnish coating layer revealed better performance of dielectric and thermal properties through BDV measurement and TGA.

Zhang et al.[36] have reviewed high-temperature resistant polymer dielectrics found applications in electrical and electronic insulating fields due to their excellent thermal stability at elevated temperatures, good dielectric properties over wide temperature and frequency ranges, good mechanical properties, and high environmental stability, etc. They included imide polymers (polyimide (PI), PAI, and PEI), polybenzimidazole (PBI), polybenzoxazole (PBO), PEEK, and polyphenylquinoxaline (PPQ).

Yang et al.[37] investigated the material compatibility between different types of lubricants and insulation materials (MW and insulation paper). They found that traditional ATF has poor material compatibility with MW overall, compared to fluids formulated for electrified transmission without sulphur components, and additives

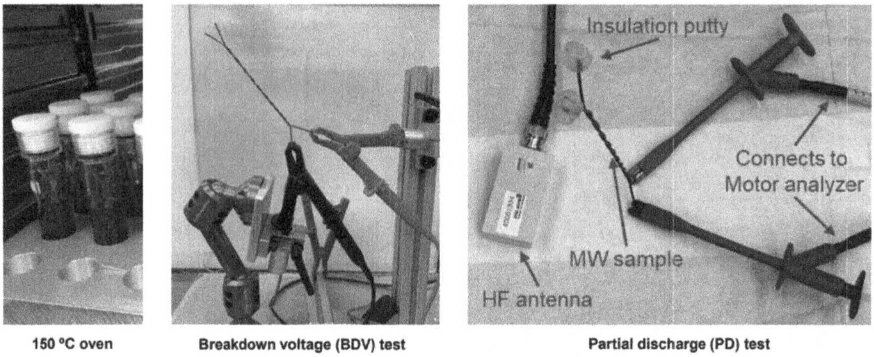

FIGURE 4.7 MW test set up with BDV and partial discharge tests.[39,41] (Used with permission of SAE International, from.[41])

play a crucial role. These were confirmed by capacitance, partial discharge inception voltage (PDIV), and peel strength of MWs.

Kwak et al.[38,39,40] explored MW compatibility studying compatibility of various ETFs and various lubricant additives per modified NEMA MW 1000 test procedures. Figure 4.7 illustrates how to test the compatibility of MW. This process typically involves aging the MW in a flask at certain temperatures for a specific duration. Temperatures can vary widely, up to about 180°C. However, it is essential to consider that too high a temperature can alter the chemistry under investigation, while too low a temperature might not reveal any significant compatibility changes for extended periods. Therefore, finding a reasonably good test condition is beneficial for efficiently screening MW compatibility with lubricants or lubricant additives.

When MW is manufactured, various performance tests are conducted to ensure its proper functionality. However, these tests are typically performed under thermal aging conditions without any lubricants. Therefore, it is crucial to assess MW compatibility with lubricants at an appropriate temperature range. Among the key performance tests for evaluating lubricant compatibility are BDV, partial discharge behavior, adhesion in addition to aged fluid's analysis. Sometimes, water is added to accelerate the aging process, but excessive water can alter the chemistry or failure mode under investigation. Therefore, care must be taken to avoid adding too high a concentration of water to the aging fluid, for example, approximately 5000 ppm.

Figure 4.8 below illustrates that MW with good compatibility with lubricants can maintain strong adhesion to the copper substrate. Conversely, MW with poor compatibility may experience delamination, as depicted in the figure. Magnet wire bending study, which is similar to this test, has been explored by Kim et al. as described above.[32]

The selection of lubricant chemistry, particularly the choice of additive components, is critical, as illustrated in the Figure 4.9 below. Depending on the detergent selected, the BDV of the MW can either improve or deteriorate. Although the specific structures of each detergent are not detailed, it is evident that both the concentration and the chemical nature of the detergent significantly impact MW compatibility. This finding might be applicable to any other lubricant additives. Therefore, making the right choice of additive components for the lubricant is crucial to maintain good MW compatibility.

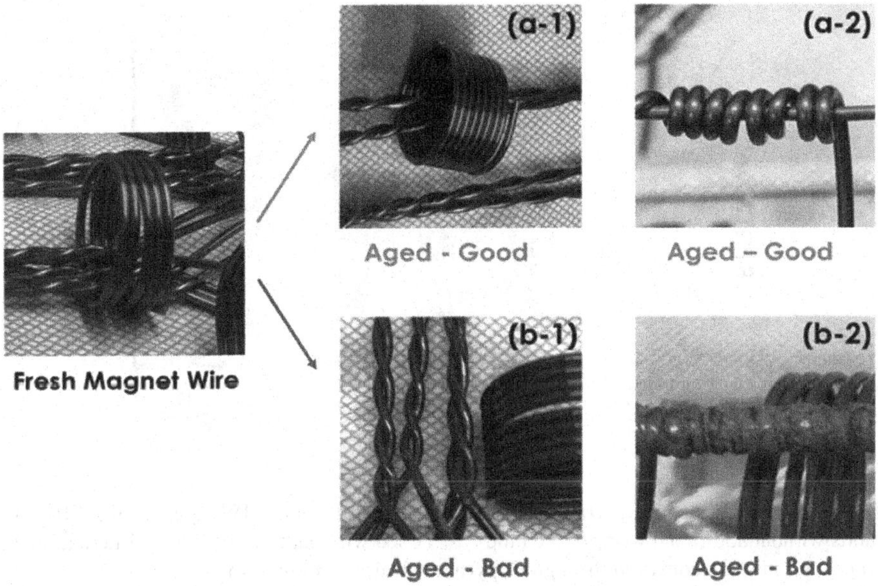

FIGURE 4.8 An example of good and poorly performing MWs in adhesion test after aging in fluids.[39,41] (Used with permission of SAE International, from.[41])

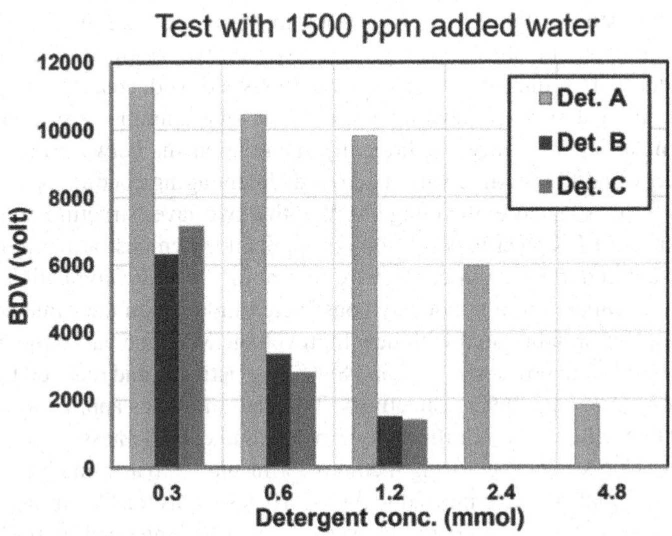

FIGURE 4.9 BDV of aged MWs in fluids with varying detergents. All tested fluids were added 1500 ppm water.[39,41] (Used with permission of SAE International, from.[41])

Figure 4.10 below demonstrates the repetitive PDIV (RPDIV). The lower the RPDIV, the poorer the performance of the MW. In the study of two fluids, Fluid B either maintained or slightly increased the RPDIV as aging time increased, when

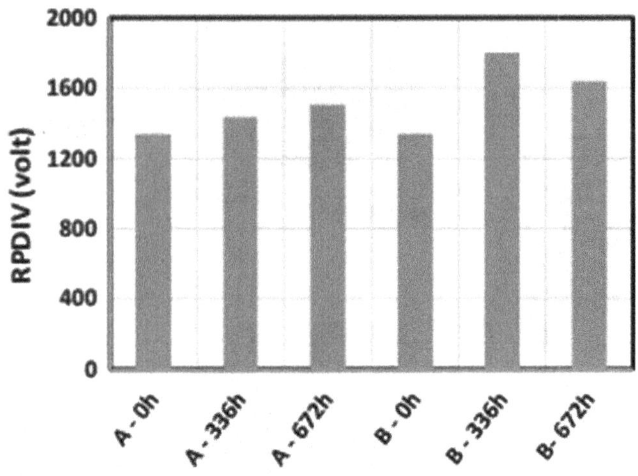

FIGURE 4.10 Repetitive partial discharge inception voltage of MWs aged in two different fluids without additional water in the fluids.[39,41] (Used with permission of SAE International, from[41]; permission conveyed through Copyright Clearance Center, Inc.)

there was no water added in the fluid. This could be due to test variation or possibly a protective layer formed by Fluid B.

Sili et al.[41] explored partial discharges in recent aircraft designs that use higher voltages. They studied the effect of temperature, pressure, and moisture on PDIV and discharge characteristics. Notably, they observed two distinct behaviors related to energy dissipation during aging: wet conditions led to decreased energy due to the formation of a semi-conducting layer, while dry conditions showed energy-dependent material endurance. Additionally, the relationship between measured dissipated energy and PD lifetime varied across different aging conditions.

Ito et al.[42] investigated high voltage MW with a two-layer structure composed of PAI layer and a PEEK layer for high voltage application demonstrating higher PDIV, longer thermal endurance, and better wire formability that are compatible in transmission oil. Compared to a previously constructed motor with the same core structure, the new motor, fabricated with new high voltage MW, increased the maximum torque and maximum power by 2.6% and 8.9%, respectively, and reduced the volume and weight by 24% and 26%, respectively. The new MW was applied to an OEM's new motor that achieved top of the class in power and compactness.

Figure 4.11 illustrates the testing methods for insulation paper and binding ropes. All plastic materials possess mechanical properties such as tensile strength, elongation, and hardness, which can degrade over time under aging conditions. Therefore, it is essential to evaluate these properties in conjunction with insulation paper, binding ropes, and other insulation materials used for the fabrication of electric motors. General procedures were followed per ASTM D638 or ISO 175 test methods after aging insulation papers in various lubricants with predetermined amounts of water in the fluids, as necessary. After aging the paper for 3 weeks at 150°C, its color generally darkened, with the paper in Fluid A darkened the most, while it remained lighter in Fluids B and C. Measurement of tensile strength revealed that Fluid B provided

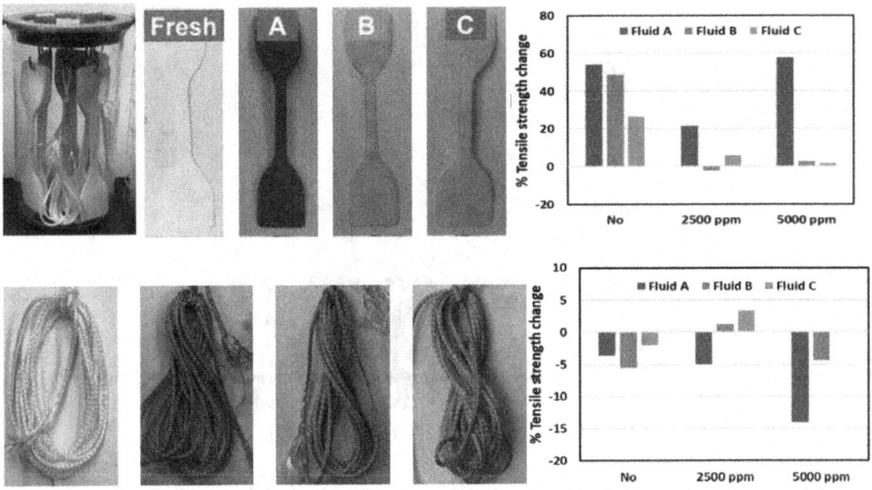

FIGURE 4.11 An example of test setup and parts (slot liner/insulation paper and binding rope) with results as a function of added water in fluids.[40] (Provided courtesy of Afton Chemical Corporation.)

the best compatibility with both insulation paper and binding ropes, as observed by the percentage change in tensile strength measured using a tensiometer. It is worth mentioning that a pressure proof vessel was used when the described amount of water was added.

Various varnish resins, such as polyester, epoxy, and PI, are available for electric motor applications, providing mechanical and electrical insulation protection to motor windings. Therefore, proper resin selection and its compatibility with lubricants are crucial. Compatibility testing can ideally be conducted within an electric motor system after varnish resin treatment, or alternatively in a laboratory using discs or other shaped samples, following appropriate curing methods. After immersion and aging in lubricants for a specified duration, parameters such as color change, tensile strength, and BDV are typically measured. Figure 4.12 illustrates an example of color changes observed in polyesterimide varnish after aging for up to 672 hours at 150°C. In shorter tests up to 168 hours, color changes were consistent across four different fluids, whereas after 672 hours of aging, Fluid B resulted in the darkest color compared to the other three fluids. Furthermore, the data demonstrate that color changes were not necessarily proportional to the sulfur content in the fluids, as depicted in the figure.

4.6 FUTURE DIRECTIONS

The importance of various materials' compatibility in lubrication environment will become more and more important going forward, particularly energy saving, efficiency, and durability perspectives. Plastics industry is driving high quality robust materials at the same time focusing on environmental aspects. Below the author outlines several research directions that both academia and industry are currently exploring and will continue to investigate further.[43]

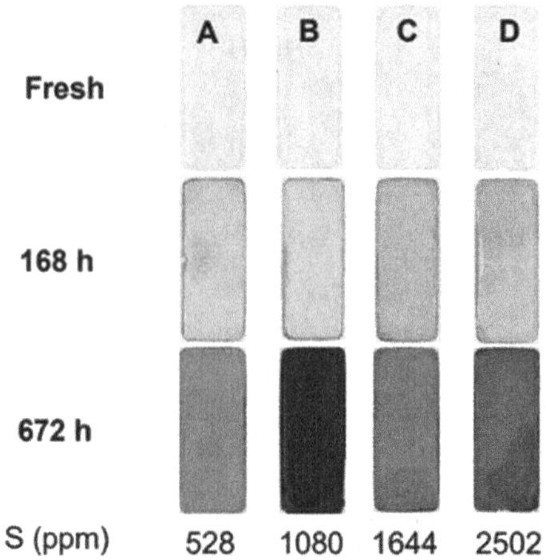

FIGURE 4.12 An example of varnish coupon compatibility. Sulphur contents on the bottom indicate ICP (inductively conductive plasma) results with each end-of-test fluid.[40] (Provided courtesy of Afton Chemical Corporation.)

- **Advanced Tribology Studies**: Investigate the tribological behavior of various engineering plastics when exposed to lubricants commonly used in EVs and HEVs. This includes assessing wear rates, friction coefficients, and surface interactions. Researchers can explore novel additives or surface modifications to enhance plastic-lubricant compatibility.
- **Surface Engineering**: Investigate surface treatments (such as coatings, plasma treatments, or nanocomposites) to enhance plastic-lubricant compatibility. Surface modifications can alter wettability, reduce friction, and improve wear resistance.
- **Modelling on Performance Prediction**: Employ multiscale modelling techniques to predict plastic-lubricant interactions at different levels: molecular, microstructural, and macroscopic. These models can guide material design, inform lubricant formulation, and aid in selecting optimal lubricant and material combinations.
- **Long-Term Durability**: Evaluate plastic materials' performance over extended periods. Accelerated aging tests can simulate real-world conditions, including temperature variations, exposure to aggressive lubricants, and mechanical stress. Understanding how plastics degrade and interact with lubricants over time is critical for ensuring vehicle reliability.
- **Recyclability and Sustainability**: Consider the environmental impact of plastic materials in EVs. Research sustainable alternatives, recycling methods, and circular economy approaches. Balancing performance with eco-friendliness is essential.
- **In Situ Monitoring**: Develop sensors or monitoring systems to assess plastic-lubricant interactions in real-time. These could detect wear, chemical changes, or degradation, allowing for proactive maintenance and material optimization.

- **Collaboration across Disciplines**: Foster collaboration between material scientists, lubricant experts, and automotive engineers. Integrating knowledge from these domains will lead to holistic solutions for plastic-lubricant compatibility challenges.

4.7 CONCLUSION

This review chapter highlights the critical importance of compatibility between non-metal materials and lubricants in the evolving landscape of electric and hybrid electric vehicle technology. Extensive analysis of various testing methodologies and compatibility challenges provides a foundational understanding for lubricant formulators, materials researchers, and application engineers in the automotive industry. Comprehensive compatibility testing, encompassing assessments of tensile and elongation strengths, crystallinity, and changes in chemical nature through various analytical tools, proves crucial for screening compatibility and advancing material and lubricant development. Also, the application of these learnings to actual production parts testing under realistic conditions should be followed to confirm compatibility in application and operation. The author emphasizes that compatibility is not merely a technical requirement but a significant factor in ensuring the durability, efficiency, and reliability of vehicle components. Looking forward, the integration of advanced materials science, innovative engineering solutions, and multidisciplinary research will be vital. This approach will facilitate the development of materials that not only meet the stringent requirements of modern automotive applications but also align with environmental sustainability goals. The continuous evolution of lubricant formulations and material engineering presents an ongoing opportunity to enhance vehicle performance and sustainability. As such, the automotive industry must remain vigilant in conducting rigorous testing and research to stay ahead of the technological curve and regulatory standards.

NOTES

1 Zhang, X.-M., Liu, J.-G., & Yang, S.-Y., "A review on recent progress of R&D for high-temperature resistant polymer dielectrics and their applications in electrical and electronic insulation," *Rev. Adv. Mater. Sci.* 46 (2016) 22–38.
2 Kondo, M. Y., Montagna, L. S., Morgado, G. F. M., Castilho, A. L. G., Batista, L. A. P. S., Botelho, E. C., Costa, M. L., Passador, F. R., Rezende, M. C., & Ribeiro, M. V., "Recent advances in the use of Polyamide-based materials for the automotive industry," *Polím. Ciência Tecnol.* 32(2) (2022) e2022023. https://doi.org/10.1590/0104-1428.20220042.
3 Gupta, P., Toksha, B., Patel, B., Rushiya, Y., Das, P., & Rahaman, M., "Recent developments and research avenues for polymers in electric vehicles," *Chem. Rec.* (2022) e202200186. https://doi.org/10.1002/tcr.202200186.
4 Tung, S. C., Woydt, M., & Shah, R., "Global insights on future trends of hybrid/EV driveline lubrication and thermal management," *Front. Mech. Eng.* 6 (2020) 571786. https://doi.org/10.3389/fmech.2020.571786.
5 Canter, N., *Tribology and Lubrication for E-Mobility: Findings from the 2nd STLE Conference on Electric Vehicles* (EV white paper), STLE, Park Ridge, IL, USA (October 2023).

6 Selema, A., Ibrahim, M. N., & Sergeant, P., "Electrical machines winding technology: Latest advancements for transportation electrification," *Machines* 10 (2022) 563. https://doi.org/10.3390/machines10070563
7 "Fuels and Lubricants TC 3 Driveline and Chassis Lubrication, *Fluid for Automotive Electrified Drivetrains*. Warrendale, PA: SAE International (2022). https://www.sae.org/standards/content/J3200_202210/.
8 ASTM D638, *Standard Test Method for Tensile Properties of Plastics*. West Conshohocken, PA: ASTM International (2014).
9 ASTM D882, *Standard Test Method for Tensile Properties of Thin Plastic Sheeting*. West Conshohocken, PA: ASTM International (2018).
10 ISO 175, *Plastics – Methods of Test for the Determination of the Effects of Immersion in Liquid Chemicals*. Geneva: ISO (2010). www.iso.org.
11 ISO 527, *Plastics – Determination of Tensile Properties*. Geneva: ISO (2012). www.iso.org.
12 ASTM D412, *Standard Test Methods for Vulcanized Rubber and Thermoplastic Elastomers – Tension*. West Conshohocken, PA: ASTM International (2021).
13 ASTM D471, *Standard Test Method for Rubber Property – Effect of Liquids*. West Conshohocken, PA: ASTM International (2021).
14 ASTM D2240, *Standard Test Method for Rubber Property – Durometer Hardness*. West Conshohocken, PA: ASTM International (2021).
15 ASTM D6515; *Standard Test Method for Rubber Shaft Seals Determination of Recovery from Bending*. West Conshohocken, PA: ASTM International (2023).
16 ASTM D7216; *Standard Test Method for Determining Automotive Engine Oil Compatibility with Typical Seal Elastomers*. West Conshohocken, PA: ASTM International (2023).
17 ISO 1817, *Rubber, Vulcanized or Thermoplastic – Determination of the Effect of Liquids*. Geneva: ISO (2024) www.iso.org.
18 CEC L-112-16, *The Evaluation of Oil - Elastomer Compatibility*. Brussels: CEC (2016). https://cectests.org/test-methods-publications
19 ASTM D149, *Dielectric Breakdown Voltage and Dielectric Strength of Solid Electrical Insulating Materials at Commercial Power Frequencies*. West Conshohocken, PA: ASTM International (2013).
20 ASTM D2519, *Bond Strength of Electrical Insulating Varnishes by the Helical Coil Test*. West Conshohocken, PA: ASTM International (2019).
21 IEC 60819-2, *Non-Cellulosic Papers for Electrical Purpose*. Geneva: IEC (2001). www.iec.ch.
22 ISO 1924-2:2008, *Paper and Board – Determination of Tensile Properties, Constant Rate of Elongation Method (20 mm/min)*. Geneva: ISO (2008). www.iso.org.
23 ANSI/NEMA MW 1000, *American National Standard for Magnet Wire*. NEMA (2021). www.nema.org.
24 ASTM D2307, *Thermal Endurance of Film-Insulated Round Magnet Wire*. West Conshohocken, PA: ASTM International (2013).
25 ASTM D1676, *Standard Test Methods for Film-Insulated Magnet Wire*. West Conshohocken, PA: ASTM International (2017).
26 IEC 60172, *Test Procedure for the Determination of the Temperature Index of Enamelled and Tape Wrapped Winding Wires*. Geneva: IEC (2015). www.iec.ch.
27 Tormos, B., Bermúdez, V., Balaguer, A., & Giménez, E., "Compatibility study of polyamide (PA6) with lubricant bases for electric vehicle applications," *Lubricants* 12 (2024) 54. https://doi.org/10.3390/lubricants12020054.
28 García, A., Valbuena, G. D., García-Tuero, A., Fernández-González, A., Viesca, J. L., & Battez, A. H., "Compatibility of automatic transmission fluids with structural polymers used in electrified transmissions," *Appl. Sci.* 12 (2022) 3608. https://doi.org/10.3390/app12073608.

29 García Tuero, A., Díez-Valbuena, G., Rivera, N., González, R., & Hernández Battez, A., "Changes in hardness and volume of polymers in contact with lubricants used in electric vehicle transmissions," In: *IACME 2022, Proceedings of the XV Ibero-American Congress of Mechanical Engineering*, Cham: Switzerland, edited by Vizán Idoipe, A., & García Prada, J. C., Springer, New York, pp. 36–42 (2023). https://doi.org/10.1007/978-3-031-38563-6.
30 Beyer, M., Brown, G., Gahagan, M., Higuchi, T., Hunt, G., Huston, M, Jayne, D., McFadden, C., Newcomb, T., Patterson, S., Prengaman, C., & Shamszad, M., "Lubricant concepts for electrified vehicle transmissions and axles," *Tribology Online*, 14(5) (2019), 428–437. https://doi.org/10.2474.trol.14.428.
31 Glasgow, M., "Lubricant additive interactions with elastomer sealing materials," In: *STLE Detroit Section Meeting*, Detroit, MI, (January 15, 2014).
32 Bennett, C., Bell, J., & Guevremont, J., "Physicochemical mechanisms for fluoroelastomer seal failures," In: *SAE Technical Paper 2016-01-2272*. Warrendale, PA: SAE International (2016). https://doi.org/10.4271/2016-01-2272.
33 García-Tuero, A., Ramajo, B., Valbuena, G. D., Fernández-González, A., Mendoza, R., García, A., & Hernández Battez, A., "Automatic transmission fluids in electrified transmissions: Compatibility with elastomers," *Appl. Sci.* 12 (2022) 6213. https://doi.org/10.3390/app12126213.
34 Norris, P., Carranza, A., Pollard, G., Gant, A., & Remias, J., "Dynamic seals for future industrial gear products," In: *Proceedings of the 2022 STLE Annual Meeting & Exhibition*, Session 3L, St. Louis, MO, USA (May, 2022).
35 Kim, S. H., & Lee, D., "Polyetherimide for MW applications," In: *SAE Technical Paper 2022-01-0334*. Warrendale, PA: SAE International (2022). https://doi.org/10.4271/2022-01-0334.
36 Zhang, X.-M., Liu, J.-G., and Yang, S.-Y., "A review on recent progress of R&D for high-temperature resistant polymer dielectrics and their applications in electrical and electronic insulation," *Rev. Adv. Mater. Sci.* 46 (2016) 22–38.
37 Yang, L., Zhang, S., Pauli, F., Charrin, C., & Hameyer, K., "Material compatibility of cooling oil and winding insulation system of electrical machines," In: *2022 International Conference on Electrical Machines (ICEM)*, Valencia, Spain, pp. 1334–1340 (2022). https://doi.org/10.1109/ICEM51905.2022.9910921.
38 Kwak, Y., Grzyska, P., Cleveland, C., & Tsuneo, A., "Magnet wire compatibility test for electric drivetrain fluid development," In: *SAE Technical Paper 2023-32-0145*. Warrendale, PA: SAE International (2023).
39 Kwak, Y., Adhvaryu, A., & Cleveland, C., "New emobility test developments and ETF technology," In: *STLE Tribology and Lubrication for E-Mobility Conference*, San Antonio, TX (November 3–5, 2021).
40 Kwak, Y., Cleveland, C., & Tsuneo, A., "magnet wire compatibility test for electric drivetrain fluid development," In: *JSAE/SAE PEL International Meeting*, Kyoto, Japan (August 29–September 1, 2023).
41 Sili, E., Cambronne, J. P., Naudé, N., & Khazaka, R., "Polyimide lifetime under partial discharge aging: Effects of temperature, pressure and humidity," *IEEE Trans. Dielectr. Electr Insul.* 20(2) (2013) 435–442.
42 Ito, K., Shibata, T., & Kawasaki, T., "Development of high voltage wire for new structure motor in full hybrid vehicle," *SAE Int. J. Alt. Power.* 5(2) (2016) 272–277. https://doi.org/10.4271/2016-01-1221.
43 Canter, N. (Ed.), *2023 Report on Emerging Issues and Trends in Tribology and Lubrication Engineering*. Park Ridge, IL: STLE (2023).

5 Introduction to Rheology for EV Fluids

Kartik Pondicherry, Paul Staudinger, and Carlos Sanchez

SYMBOLS AND TERMS

Symbol	Parameter
F	Force
A	Area
D	Gap
R	Radius
M	Torque
h	height of sample
n	Rotational speed
τ	Shear stress
γ	Shear strain
$\dot{\gamma}$	Shear rate
η	Viscosity
δ	Phase shift angle
σ_A	Stress amplitude
ε_A	Strain amplitude
G^*	Complex modulus
G'	Storage modulus
G''	Loss modulus

5.1 RHEOLOGY

Rheology is the branch of science dealing with the study of flow and deformation behavior of materials. Its etymology stems from the Greek word "rheo," which means "to flow." Although it primarily applies to liquids and semi-solids that exhibit flow behavior, rheology is also used to describe the deformation behavior of solids.

In rheology, materials can be classified into three categories: ideally viscous, like water and mineral oils; viscoelastic, which includes substances like grease, shampoo, and ketchup; and ideally elastic, such as steel and stone (Table 5.1).

5.2 VISCOSITY

Viscosity is a measure of a fluid's resistance to flow, and it describes how easily it deforms under an applied force. Viscosity of a sample is measured with the help of

DOI: 10.1201/9781003496069-5

TABLE 5.1
Categorization of Materials in Terms of Their Rheology

Ideally Viscous	Viscoelastic	Ideally Elastic
Water	Shampoo, lotion,	Steel
Mineral oil	grease, bitumen	Stone

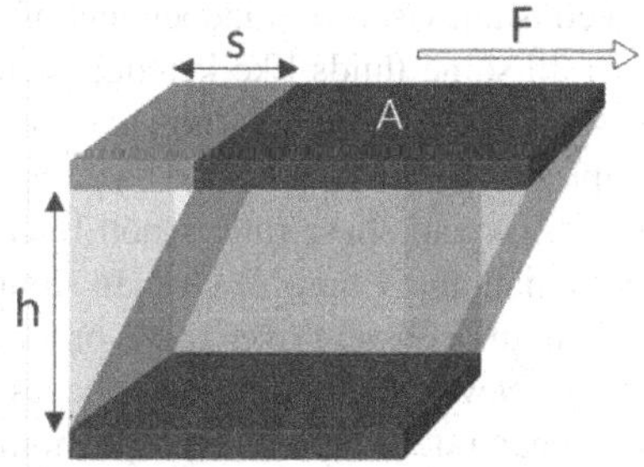

FIGURE 5.1 Sketch depicting the two-plate model. The area of the plate is A, and the height of the sample is h. The displacement at the topmost layer is depicted by s and F is the applied force.

the two-plate model, which in turn uses Newton's law pertaining to the application of force. As illustrated in Figure 5.1, the fluid is sandwiched between two parallel plates. The basic assumption in this case is that the sample comprises infinite parallel laminar layers. When a force F is applied to the upper plate to make it move at a constant velocity, it also drags the fluid layer below it due to friction. This dragging effect is applied to the lower layers of fluid as well, and the extent of flow depends upon the internal friction, or the viscosity of the fluid.

To mathematically define viscosity, it is important to introduce two additional parameters: shear stress (τ) and shear rate ($\dot{\gamma}$). Shear stress is defined as force per unit area applied to the upper plate. Shear rate is the velocity gradient across the fluid layer; the ratio between the applied force F and the area A.

$$\tau = \frac{F}{A} \tag{5.1}$$

where τ is the shear stress, F the applied force, and A the area of the plate.

$$\dot{\gamma} = \frac{v}{h} \tag{5.2}$$

where $\dot{\gamma}$ is the shear rate, v the velocity or the rate of change of displacement, and h the distance between the upper and the lower plate.

As per Newtons law, the shear rate is directly proportional to the applied shear stress and the proportionality constant is referred to as the viscosity of the fluid.

$$\tau = \eta \, \dot{\gamma} \tag{5.3}$$

$$\eta = \frac{\dot{\gamma}}{\tau} \tag{5.4}$$

where $\dot{\gamma}$ is the shear rate, τ the shear stress, and η the viscosity of the fluid.

5.3 NEWTONIAN AND NON-NEWTONIAN FLUIDS

In fluids like water or mineral oil, the relationship between shear stress and shear rate is linear, i.e., the fluid has a constant viscosity independent of the shear rate. These are known as Newtonian fluids. In some fluids like ketchup or toothpaste, the viscosity changes depending on the shear rate. These are known as non-Newtonian fluids. The flow behavior of such samples cannot be described by a constant viscosity, and the relationship between shear stress and shear rate is non-linear. Figure 5.2 illustrates the flow behavior of a Newtonian and a non-Newtonian sample. While the viscosity of the Newtonian sample remains constant over three orders of magnitude of shear rate, the viscosity of the non-Newtonian sample decreases by at least two orders of magnitude in the same shear rate window. This phenomenon is also known as shear-thinning.

5.4 VISCOELASTICITY

Materials like water or oils display ideally viscous behavior, where any applied stress is fully absorbed to cause the fluid to flow or deform without storing any energy.

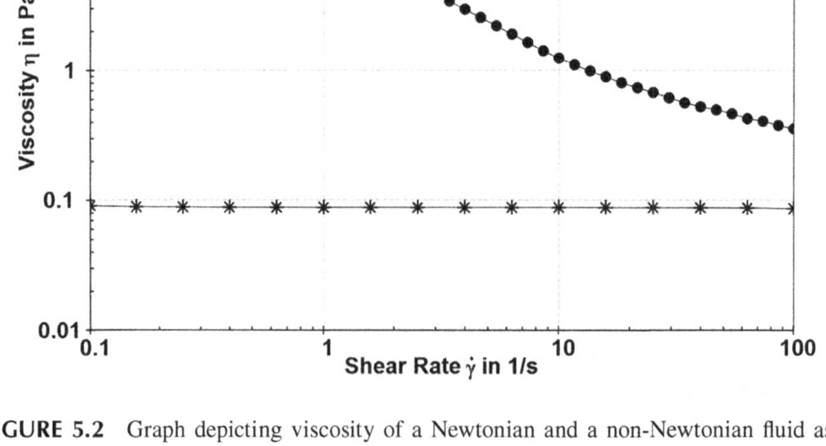

FIGURE 5.2 Graph depicting viscosity of a Newtonian and a non-Newtonian fluid as a function of shear rate.

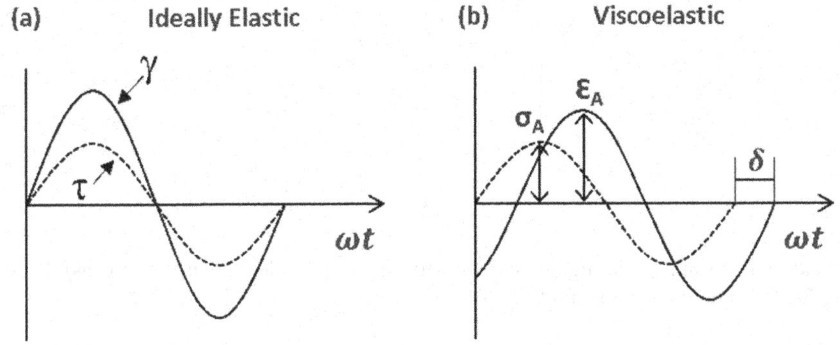

FIGURE 5.3 Plots depicting the strain response of (a) ideally elastic sample, and (b) a viscoelastic sample, to the applied stress.

Viscoelastic materials, such as greases and shampoo, exhibit a blend of both viscous and elastic properties when they undergo deformation.

Viscoelastic properties of a sample can be measured using oscillatory rheometry, where the sample is exposed to cyclic stresses or strains, and its response is monitored. Figure 5.3 illustrates the strain (γ) response of a sample when subjected to a sinusoidal stress (τ). In the case of an ideally elastic material, such as steel, the strain is perfectly in phase with the applied stress. This means that the strain signal aligns exactly with the stress signal, indicating that the material instantaneously deforms and recovers when the stress is applied and removed. There is no time delay between stress and strain in an ideally elastic material. In the case of viscoelastic materials, the strain signal lags behind the applied stress, indicating a phase shift between stress and strain. This phase difference reflects the material's combination of elastic and viscous behavior, where some of the energy is stored elastically and some is dissipated as viscous flow. The phase difference or phase shift angle δ indicates the lag between the applied stress and the resulting strain. For ideally elastic materials, $\delta = 0°$, for ideally viscous materials, $\delta = 90°$.

5.5 DYNAMIC MODULI

The complex modulus G^*, as shown in Equation 5.5, represents the stiffness of the sample. The storage (G') and loss (G'') moduli of the sample are derived from the complex modulus and the phase shift angle δ as shown in Equations (5.6) and (5.7). The storage modulus (G') reflects the elastic characteristics of the sample, while the loss modulus (G'') depicts its viscous characteristics. The relationship between the complex modulus, loss modulus, and storage modulus is depicted in Figure 5.4. As mentioned in the previous section, the phase shift angle δ ranges from 0° to 90°. The smaller the angle, the more elastic the sample is, and with increasing angle, the viscous characteristics also increase.

$$G^* = \sigma A / \varepsilon A \tag{5.5}$$

where G^* is the complex modulus, σA is the stress amplitude, and εA is the strain amplitude, see Figure 5.3b

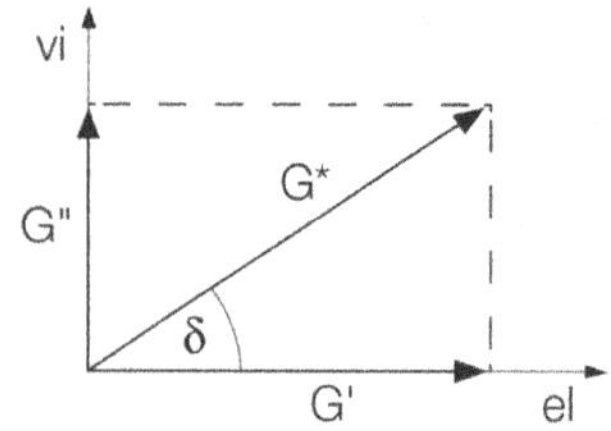

FIGURE 5.4 Graph depicting the relation between the complex modulus (G^*), and the loss modulus (G'') and the storage modulus (G').

$$G' = G^* \cos \delta \tag{5.6}$$

$$G'' = G^* \sin \delta \tag{5.7}$$

where G' is the storage modulus and, G'' is the loss modulus.

5.6 RHEOMETRY

Rheological characteristics can be studied with the help of scientific instruments known as rheometers. While rotational and oscillatory rheometers are very common, there are also other types of rheometers such as capillary, falling body, and extensional rheometers. The choice of rheometer depends on the parameters that one intends to study as well as the properties of the samples. In this chapter, the focus is only on rheological measurements carried out on a rheometer with rotational and oscillation capabilities. A simplified sketch of a rheometer is shown in Figure 5.5. While rotational measurements are used to study the viscosity of the materials under different mechanical and environmental conditions, oscillatory measurements offer information regarding viscoelastic properties of the samples. The following sections provide more detailed information regarding these testing procedures.

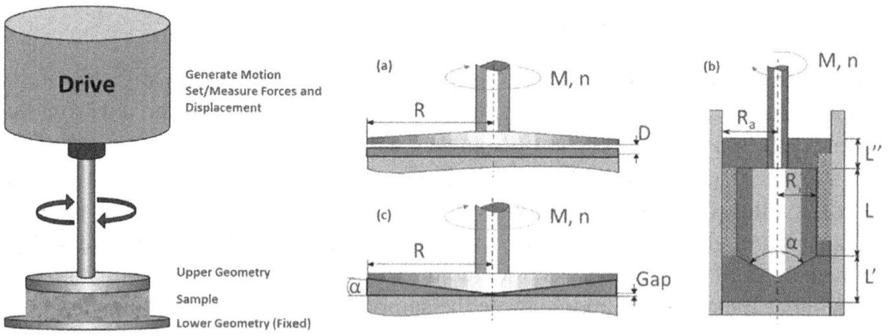

FIGURE 5.5 Basic schematic of the principle of rotational/oscillatory rheometry along with the most commonly used measurement geometries: (a) parallel-plate, (b) concentric cylinder, and (c) cone-plate system.

5.7 AMPLITUDE SWEEP

In an oscillatory test, the stress or strain input is in the form of a sinusoidal wave, and such waves are characterized by their amplitude and frequency. In an amplitude sweep, as the name suggests, the amplitude of the set shear stress or shear strain is gradually (logarithmically) increased while keeping the frequency constant. The aim of this test is to measure the sample's response to the increasing shear and obtain information pertaining to its viscoelastic properties. The data obtained from an amplitude sweep is typically depicted in the form of a graph showing G' and G'' as a function of either the shear strain or the shear stress as shown in Figure 5.6a and b, respectively. Such a depiction makes it easy to read some basic information from the graph. For instance, in this case, as the value of G' is greater than that of G'' in the plateau regime, the sample can be classified as a viscoelastic solid. The plateau region of the storage modulus indicates that the relationship between stress and strain is linear, and this is known as the linear viscoelastic region (LVER). Typically, this region extends to the point where the value of G' drops by 3% or 5%. Up to this point, the internal structure of the sample remains undisturbed. This is also known as the yield point, beyond which the sample shows non-linear behavior and the material can no longer retain its original shape after deformation. The point where the G' and G'' meet is known as the crossover point, or flow point. Beyond this point, the sample transforms into a viscoelastic liquid where G'' is greater than G'.

Amplitude sweeps are also used as a precursor test for other oscillatory tests such as frequency sweeps, time sweeps, temperature sweeps, etc. As mentioned earlier, oscillatory tests require an amplitude and a frequency. Most often, these oscillatory tests are carried out within the linear viscoelastic region. These tests can also be run at various temperatures to observe the influence of thermal stress. Materials such as greases are often studied using oscillatory rheometry.

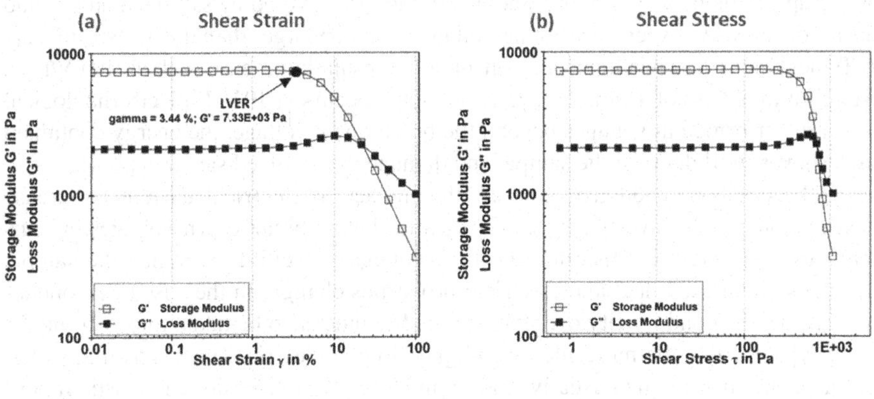

FIGURE 5.6 Graph depicting data from an amplitude sweep test. Loss and storage moduli as a function of (a) shear strain and (b) shear stress.

5.8 ELECTRORHEOLOGY

In addition to mechanical stress and temperature, electrical fields can also influence the behavior of viscoelastic materials, such as lubricating greases. Rheology testing performed under an electric field is known as electrorheology. The change of mechanical properties such as viscosity and modulus or electrical properties, such as conductivity, also has an impact on the service life of a lubricant.

There are two basic examination methods: Impedance spectroscopy and electrorheological measurements. Impedance spectroscopy aims to measure the electrical properties without changing the sample to gain insight into the structure of a sample, while in electrorheological measurements the electric field is specifically applied to influence sample properties.

5.9 INTRODUCTION TO FLUID POLARITY

Many fluids will consist of polar molecules, which are affected by electric field. Material properties such as viscosity, modulus, or conductivity can change in the presence of these fields. Electro-rheological fluids (ERF) are suspensions of electrically polarizable particles in a non-conducting liquid. They exhibit fast, strong, and reversible changes in their rheological properties upon application of an electric field. These materials are commonly used in applications such as fast responding adjustable dampers or in clutches for cars [1].

The breakdown field strength or dielectric strength (also break down voltage) indicates the highest field strength that can be applied to a material without it failing as an insulating medium. At voltages beyond the break down voltage, a flashover occurs. The break down voltage of air is about 3 kV/mm, mainly depending on humidity.

Additional measures have to be taken in a parallel plate-plate setup to apply voltages beyond the break down voltage of air (e.g., cover the free sample surface with low viscous and high dielectric strength insulating oil). Alternatively, a concentric cylinder setup can be used, as the gap and its vicinity are covered by sample material and the air distances between parts under voltage are much larger than the measuring gap. A Typical break down voltage of a conventional insulating grease is about 10 kV/mm. At a film thickness of 1 μm, voltage break down occurs at 10 V. If electrorheological tests are performed at voltages beyond the break down voltage, the energy contained in flashovers will degrade the sample and damage the involved surfaces [2–4].

In the case of conductive greases, the impact of electrical current on grease degradation can be investigated at voltages far below break down voltages, as currents are dissipated and therefore no high voltage can build up. When the sample degrades, the mechanical and electrical properties change. In the case of a conductive grease, the conductivity over the service life must also be taken into account.

In typical ERF, the molecules are aligned by the electrical field, forming superordinate structures which usually lead to an increase in viscosity and elastic modulus (Figure 5.7) [5]. The application of an electric field usually leads to a change in viscosity [6].

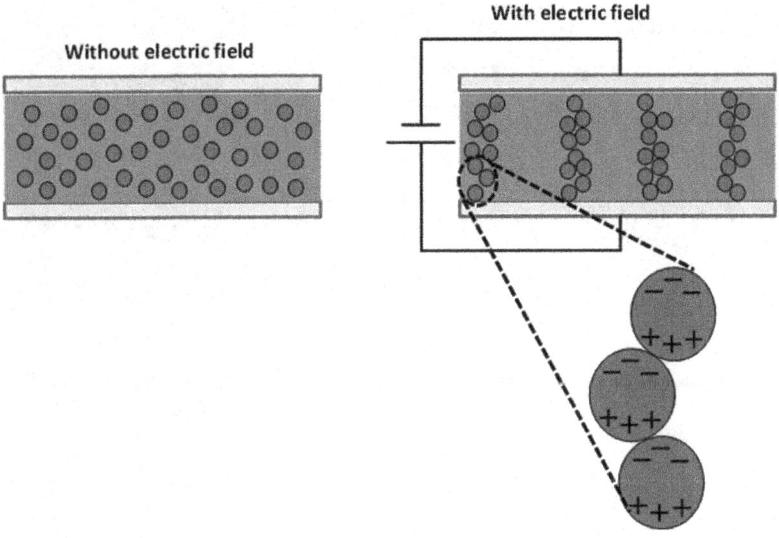

FIGURE 5.7 Alignment of ERF molecules in an electric field.

5.10 TEST TYPES

A rotation and oscillatory type rheometer can be used to perform the following types of tests to investigate the behavior of greases and other fluids.

- **Rotation**: Viscosity as function of applied electrical field (DC or AC)
  - Constant electric field strength and changing shear rate (Flow curve)
  - Constant shear rate and changing electric field strength (Voltage sweep)
- **Oscillation**: Complex viscosity, elastic and viscous moduli as function of applied electrical field
  - Constant electric field strength and changing deformation or frequency (amplitude sweep (AS), frequency sweep (FS))
  - Constant deformation and changing electric field strength (Voltage sweep)

A few example tests with results discussion are presented in the next section.

5.11 IMPACT OF ELECTRIC FIELD ON VISCOSITY

A rotational viscosity test was performed on an electro-rheological fluid (ERF) at constant shear rate in intervals from 0.1 to 1000 1/s, and constant temperature. At each constant shear test, the applied voltage was ramped from 0 to 2500 V DC. The lowest shear rate has the highest viscosity, while the highest shear rate has the lowest viscosity. These results indicate a shear-thinning behavior. At each constant shear curve, the viscosity increases with increasing voltage. The higher the applied

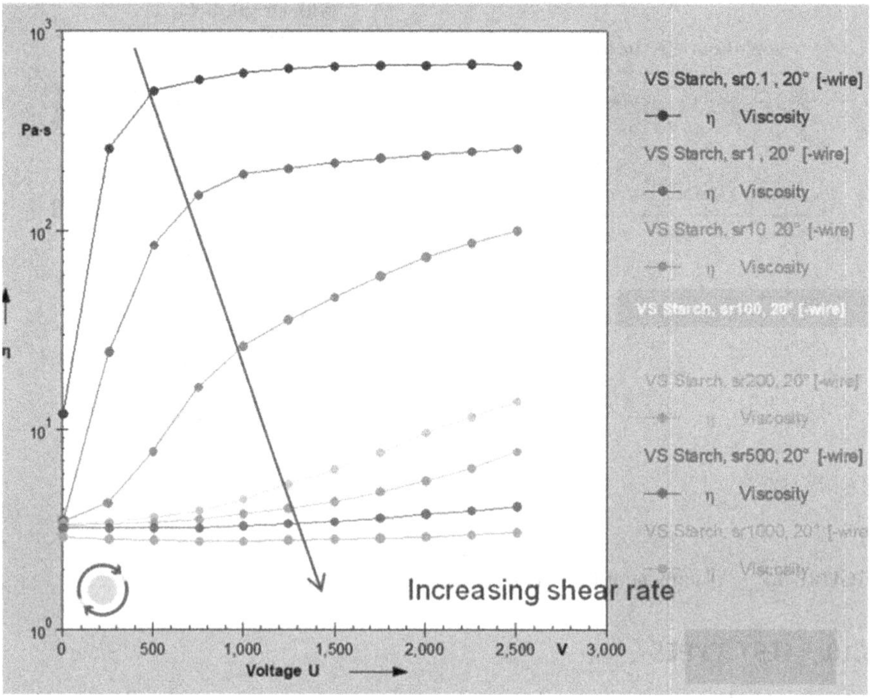

FIGURE 5.8 Constant shear rate and changing electric field strength.

voltage, the more the particles align to form a stronger structure that resists flow. Higher shear rates disturb the alignment of particles, and the influence of voltage is less. Results are shown in Figure 5.8 [5].

Another rotational viscosity test was performed on an ERF at constant voltage in intervals from 100 to 4000 V DC, and constant temperature. In this experiment, the shear rate was increased from 0.1 to 1000 1/s. The lowest viscosity was observed at the lowest voltage, while the highest viscosity was observed at the highest voltage. In all cases, the viscosity decreased with increasing shear rate, indicating a shear-thinning fluid. As in the previous case, the higher the applied voltage, the more the particles aligned to form a stronger structure that resisted flow. Similarly, higher shear rates disturbed the particle formation, thereby reducing viscosity. Results are shown in Figure 5.9.

5.12 IMPACT OF ELECTRIC FIELD ON VISCOELASTICITY

An oscillating parallel plate test was performed on an ERF. The test was an amplitude sweep at constant frequency, variable strain, and constant temperature. The applied voltage was varied for each test at intervals from 0 to 2000 V DC. The storage modulus (G') and loss modulus (G'') were measured and plotted. The fluid became more

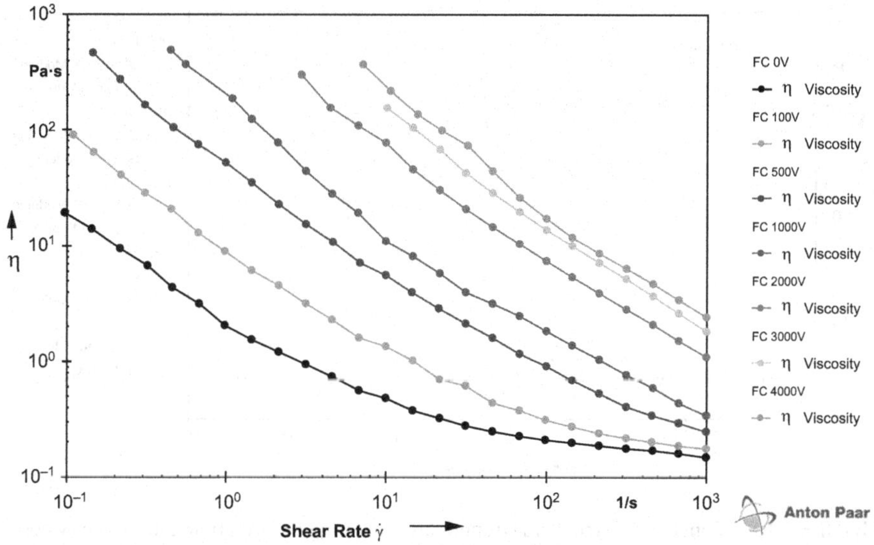

FIGURE 5.9 Constant electric field strength and changing shear rate (flow curve). *Note*: The line curve for FC 0 V is the lowest and closest to the *x*-axis, while the FC 4000 V line curve is the highest and farthest vertically from the *x*-axis.

elastic with increasing voltage. The higher the applied voltage, the more the particles in the fluid formed a structure to resist flow. At high voltages of 1000 and 2000 V, the fluid appears to be saturated to about 1% strain, though the structural strength continues to increase at higher strain rates. Results are shown in Figure 5.10.

An oscillating test was performed on a conductive grease with ionic fluid. The test was constant deformation amplitude and constant temperature. The strain values were 0.05%, 10%, and 100%. For each constant strain test, the voltage was increased from 0 to 4500 V. The linear viscoelastic (LVE) range was determined for each test. In all cases, measurement stopped once voltage breakdown occurred. The electric field was shown to have an impact on viscoelastic behavior. The peaks and inflections of the G' and G'' curves indicate a structural change in the grease. Results are shown in Figure 5.11.

An oscillating test was performed on a conventional moly-based grease at variable temperature, with and without DC current. The test was a series of constant strain frequency sweeps at increasing temperature from 16°C to 120°C at 2°C increments. The moduli for each test were plotted, and a transformation was performed at constant frequency of 10 Hz. The tan delta (loss factor) for the grease with and without 30 V 2 A DC was plotted. The applied electric field had an impact on the elasticity of the grease, causing a significant decrease in loss factor for all temperatures. The results are shown in Figure 5.12.

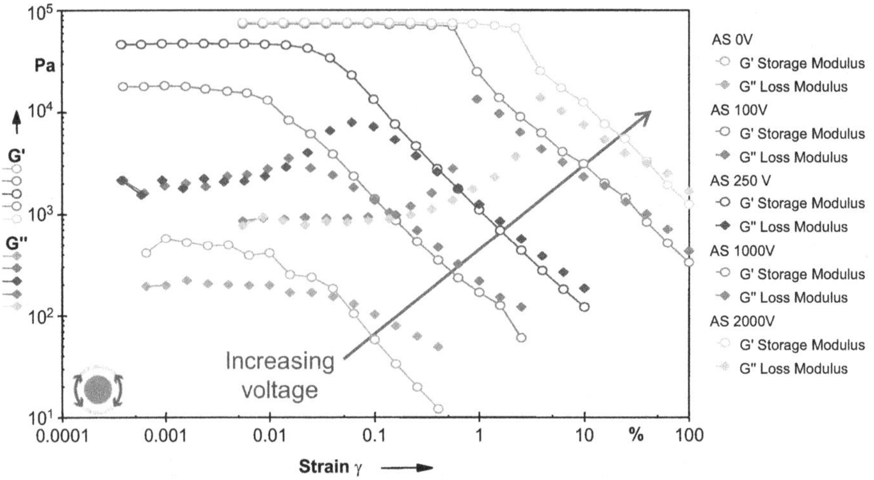

FIGURE 5.10 Constant electric field strength and changing deformation (amplitude sweep).

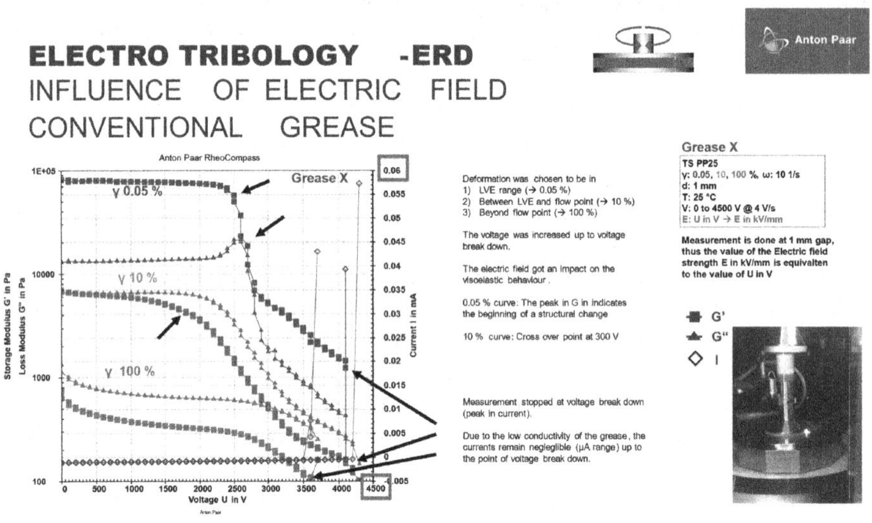

FIGURE 5.11 Storage and loss moduli and changing electric field in conventional grease.

5.13 CONCLUSIONS

Rheology testing offers many options to study fluids in ways that replicate real-world conditions. Controlled stresses such as strain and temperature can be applied to better understand fluid response. In addition to mechanical and thermal stress, electrification can also be introduced to replicate the effects seen in electric vehicle applications. Experimentation has shown that many fluids will be affected by an electric field. The viscosity may increase or decrease depending on the chemical

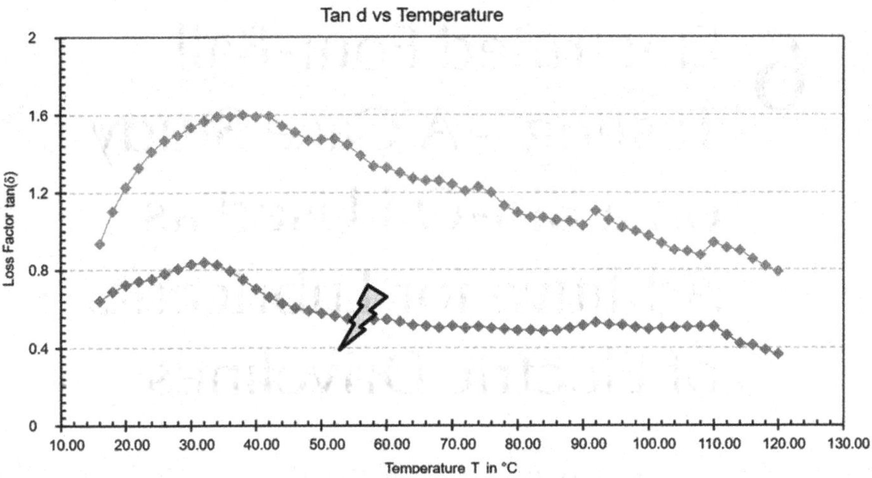

FIGURE 5.12 Loss factor in temperature sweep at constant deformation with and without applied DC voltage.

composition. Likewise, the stiffness or elasticity of a material may change. The test geometry and parameters used to study these materials can be chosen based on the working conditions. In some cases, multiple test methods may be needed to observe the changes in fluid properties.

REFERENCES

1. Jadhav, A., and Gawade, S. S. (2014). Modern applications of electrorheological (ER) fluids. *International Journal of Emerging Technologies and Applications in Engineering, Technology and Sciences (IJ-ETA-ETS)*, 1, 112–118.
2. Wang, X., and Wang, Z. (2008). Particle effect on breakdown voltage of mineral and ester based transformer oils. In: *2008 Annual Report Conference on Electrical Insulation and Dielectric Phenomena*, IEEE, Quebec, QC. doi:10.1109/CEIDP.2008.4772859.
3. Tischmacher, H., Gattermann, S., Kriese, M., and Wittek, E. (2010). Bearing wear caused by converter-induced bearing currents. In: *IECON 2010* – 36th *Annual Conference on IEEE Industrial Electronics Society*, Glendale, CA. doi:10.1109/IECON.2010.5675212.
4. Gunderson, S., Fultz, G., Snyder, C., Wright, J., Gschwender, L., and Heidger, S. (2011). The effect of water content on the dielectric strength of polyalphaolefin (PAO) coolants. *IEEE Transactions on Dielectrics and Electrical Insulation*, 18, 295–302. doi:10.1109/TDEI.2011.5704521.
5. Méheust, Y., Parmar, K. P. S., Schjelderupsen, B., and Fossum, J. O. (2011). The electrorheology of suspensions of Na-fluorohectorite clay in silicone oil. *Journal of Rheology*, 55(4), 809–833. doi:10.1122/1.3579189
6. Sanchez, E. (2023). Electro-Rheology. In: 2023 *Tribology and Lubrication for E-Mobility: Findings from the 2nd STLE Conference on Electric Vehicles*, STLE, Chicago, IL.

6 Electrified Four-Ball Testing – A Case Study on a Bio-Oil Used as Additive for Lubricants of Electric Drivelines

Carlos C. Rubio-Hernández, Peter Lee, Carlos Sanchez, Ali Erdemir, Cesar D. Resendiz-Calderon, and Leonardo Israel Farfan Cabrera

6.1 INTRODUCTION

Electric vehicles (EVs) require new low-viscosity lubricants that are able to keep or enhance the performance of the drivetrain. In contrast to traditional engine or transmission lubricants, electric drivetrain lubricants must be reformulated to provide different properties related to thermal conductivity, material compatibility, antifoaming, etc. [1,2]. Besides, they must have appropriate dielectric strength and good lubricity under the presence of shaft currents/voltages, both in order to minimize the damaging effect of shaft currents arcing to the interfaces of moving elements like bearings, gears, etc. [3–5]. Considering the global environmental protection and sustainability demands for different industrial sectors, including the automotive, the eco-friendliness or biodegradability of lubricants for EVs must also be prioritized for the new lubricant formulations.

According to a wide number of literature reports compiled in many comprehensive reviews [6–10], vegetable oils (VOs) have been stated as the best option for the development of renewable and biodegradable lubricants (bio-lubricants) for different industrial applications, including transportation vehicles [8]. VOs are composed of about 98% fatty acids promoting several convenient properties such as high flash point, very low coefficient of friction (CoF), low-to-medium viscosity ranges, and high viscosity index [7]. Bio-lubricants can be produced through neat vegetable oils [7], fatty esters [11] or estolides [12] either as base oils or as additives for mineral- or synthetic-base oils. The addition of VOs into mineral- or synthetic-base oils is, in general, helpful to enhance lubricity while improving biodegradability rates [13].

 DOI: 10.1201/9781003496069-6

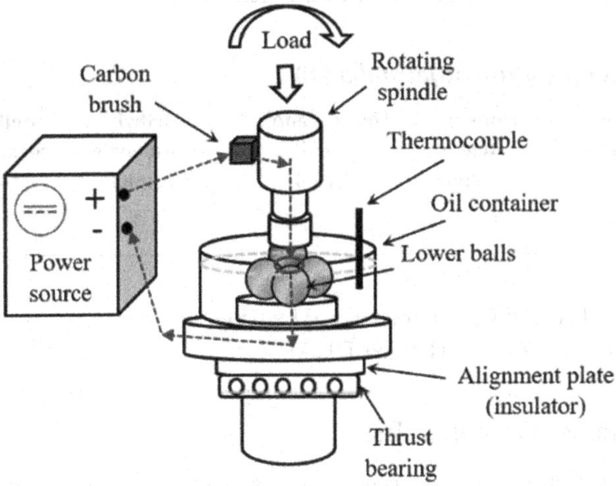

FIGURE 6.1 Schematic diagram of the electrified four-ball test set-up.

As a reference of the current progress and acceptability of bio-lubricants, global companies such as BP PLC, Chevron Corp, ExxonMobil, Repsol, Royal Dutch Shell PLC, Total, and FUCHS, among others, are already producing and marketing different green-labeled bio-lubricants for a wide range of applications [14].

Nonetheless, despite that many bio-lubricants have been widely demonstrated and recognized to exhibit good lubricity properties for numerous applications, their behavior under the specific operating conditions in the EV drivelines has not yet been explored nor reported before. In particular, the tribological performance of lubricants under electric and/or magnetic fields is critical to have an insight into the performance of the lubricant in electrified tribological systems [5], but it is also scarce in the literature. Thus, this work aims to contribute with a pioneering evaluation on the tribological and rheological properties of bio-lubricants made of blends of a mineral base oil (Group II) and a non-edible *Jatropha* oil under electrified conditions (0, 1.5, and 3 A, DC) (see Figure 6.1). *Jatropha* oil is considered one of the most favorable bio-oils for producing industrial and automotive lubricants due to its very low unsaturation promoting moderate thermal stability, high viscosity index, and good lubricity [15,16], as well as the fact that it is compatible with most of the elastomeric materials used for automotive seals [17] in comparison to other VOs.

6.2 EXPERIMENTAL DETAILS

6.2.1 Oil Samples Preparation

The oils tested in our study were straight *Jatropha* oil, a Group II-110 N mineral base oil (Group II), and two blends of Group II and Jatropha oil (5 v/v% and 10 v/v%), which will be named "Blend 5% JO" and "Blend 10% JO", respectively. To maintain the raw low-unsaturation degree, the *Jatropha* oil was not subjected to any refining or chemical conversion process. The fatty acids profile of *Jatropha* oil is

TABLE 6.1
Fatty Acids Composition of *Jatropha* Oil

Fatty Acid	Oleic Acid C18:1n9c	Linoleic Acid C18:2n6c	Hexadecanoic Acid C16:0	Methyl Stearidonate C22:6n3	Methyl Stearate C18:0	Others
%	42.2	17.5	13	11.5	8	7.8

presented in Table 6.1. It was characterized by ESI mass spectroscopy analyses using a MALDI-TOF spectrometer (Bruker DHAP).

6.2.2 Electrified Four-Ball Test

An instrumented four-ball tester was used for conducting both unelectrified and electrified tribological (CoF and wear scar diameter) evaluations of the oils according to the test conditions given in Table 6.2. The test conditions were based on the standard ASTM D4172 procedure and electrified conditions according to the method reported in [18]. A DC current power supply (Keithley® 2230-30-1) was instrumented into the four-ball tester to provide continuous electricity passing through the four balls contact and keeping the contact insulated. A schematic view of the set-up is shown in Figure 6.1. The tested balls were standard AISI 52100 bearing steel balls with hardness of 64 HRC, 12.7 mm in diameter and a surface roughness of 0.02 μm. CoF was measured and logged by the tester´s software for post analyses. The resulting wear scars on the three lower balls were measured and analyzed by optical microscopy (Keyence, VHX).

6.2.3 Electrified Rheological Test

To determine possible changes in the oil viscosity due to the applied electrical fields, the shear viscosity of the oils was determined by using a rotational rheometer

TABLE 6.2
Four-Ball Testing Conditions

Test Parameter	Values
Load [N]	392
Maximum Hertzian contact pressure [GPa]	4.5
Speed [rpm]	1200
Test duration [minutes]	60
Temperature [°C]	75 ± 2
Humidity [%]	35
Amount of oil [mL]	30
(DC) current [A]	0, 1.5, 3
Repeats (using new oil sample and new balls)	≥ 3

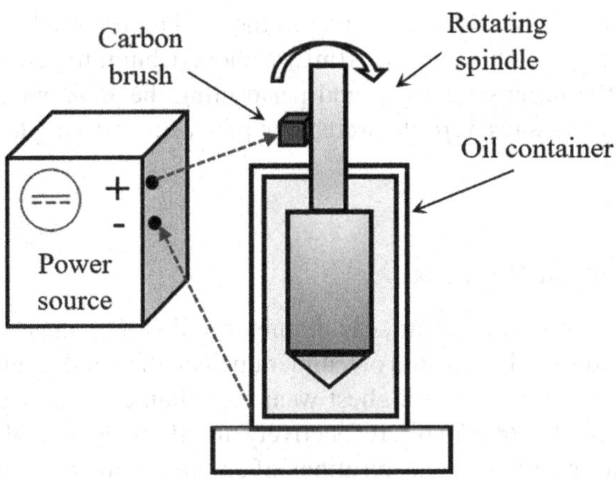

FIGURE 6.2 Schematic diagram of the electrified concentric cylinders rheometer.

(Anton Paar MCR702e Multidrive) equipped with an electro-rheological cell coupled to a concentric cylinders flow geometry (DG28.91), as shown in Figure 6.2. This device and geometry were appropriate for measuring the shear viscosity of the low-viscosity samples at 20°C and 75°C under electrical fields with 0, 1.5 and 3 A, respectively, in shear rate ramps from 500 to 1000/s.

6.2.4 Dielectric Strength Test

Dielectric strength or breakdown voltage (BDV) of oils represents their insulation capacity and provides an insight into electrical conductivity, which is relevant for the tribological behavior of oils subjected to electrical environments [4,11]. The BDV tests were carried out in an insulation oil tester (kv tester, ZC-210B) with flat electrodes under the ASTM D877 conditions for in-use electrical liquids, as listed in Table 6.3. The BDV was measured with DC at different temperatures (from 30°C

TABLE 6.3
Breakdown Voltage (BDV) Test Conditions

Parameter	Values
Disc gap (mm)	2.5 ± 0.01
Electrode type	Flat (3 mm thickness)
Electrode material	Brass
Fluid quantity (mL)	250 mL
Input voltage (V)	AC 220
Output voltage (kV)	0–80
Voltage boosting speed (kV/s)	2% ± 10%
BDV measurement repeats	Five per each oil sample
Oil temperature (°C)	30°C, 40°C, 50°C, 60°C, 70°C

to 70°C with intervals of 10°C) according to the method reported in [19]. The tests consisted of pouring an oil sample (250 mL) without dehumidification at the desired temperature in the tester´s fluid cup and performing the BDV measurement. Five BDV measurements, as test repeats, were done per each fluid sample.

6.3 RESULTS

6.3.1 Tribological Behavior

Figure 6.3a and b show the wear scar diameter and CoF change results, respectively, determined for all the tested oils under unelectrified and electrified four-ball test conditions. The lowest and highest wear scar diameter and CoF were generated by Group II and *Jatropha* oil, respectively, for all unelectrified and electrified cases. The addition of both concentrations of *Jatropha* oil in Group II promoted significant reductions of wear and CoF for all the tested conditions. The applied electricity generated an increase in wear but a decrease in CoF for all the oils, in particular at 3 A. Figures 6.4–6.6 show examples of the wear scars produced by using Group II, *Jatropha* oil, and Blend 10% JO at the different electrical currents applied. In all cases, deep abrasion grooves, especially at the center of the scars, are the predominant wear patterns. However, in both electrified Group II cases (see Figure 6.4b and 6.c), some regions with scuffing can be seen.

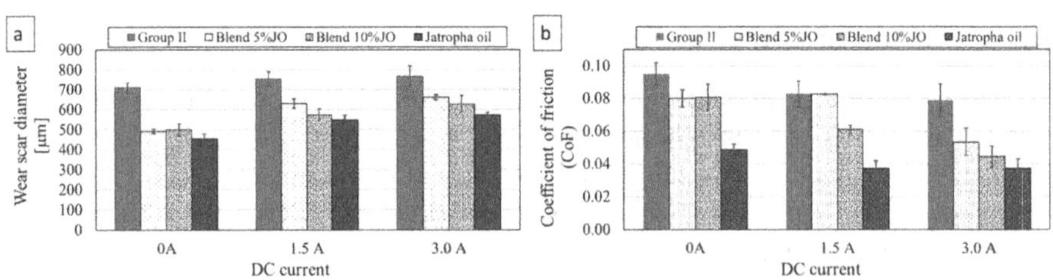

FIGURE 6.3 Tribological results obtained for the oils under non-electric (0 A) and electric (1.5 and 3 A) four-ball environments: (a) wear scar diameter; (b) coefficient of friction (CoF).

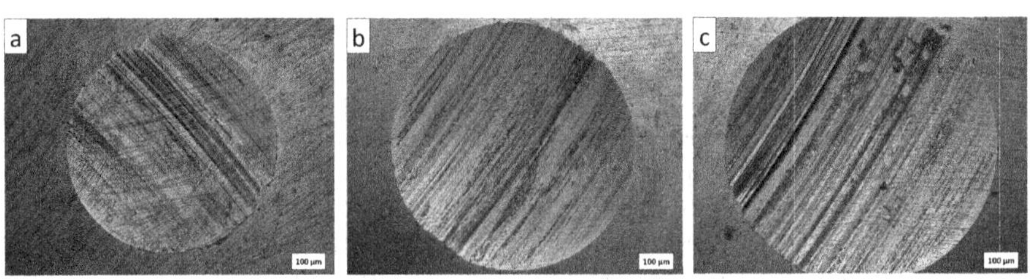

FIGURE 6.4 Wear scars generated by using Group II for the different electrical conditions: (a) 0 A; (b) 1.5 A; (c) 3 A.

FIGURE 6.5 Wear scars generated by using Jatropha oil for the different electrical conditions: (a) 0 A; (b) 1.5 A; (c) 3 A.

FIGURE 6.6 Wear scars generated by using blend 10% JO for the different electrical conditions: (a) 0 A; (b) 1.5 A; (c) 3 A.

6.3.2 Viscosity

The results of shear viscosity (from 500 to 1000^{-1} shear rates) for all the oils exhibited a Newtonian behavior; the shear-viscosity did not change with shear rate for any condition. Comparisons of the mean shear viscosity obtained for all the oils in the range of 500 to 1000^{-1} at 20°C and 75°C are shown in Figure 6.7a and b, respectively. At both temperatures, any of the applied DC currents (1.5 and 3 A) generated meaningful changes in shear viscosity for any oil. Furthermore, Group II presented the lowest viscosity values (41 and 7.2 mPa·s at 20°C and 75°C, respectively) while *Jatropha* oil presented the highest viscosity values (66.9 and 13.8 mPa·s at 20°C and 75°C, respectively). However, both additions of *Jatropha* oil (5 and 10 v/v.) in Group II were found to generate a minimal increase of shear-viscosity in the order of 1 or 2 mPa·s at both temperatures.

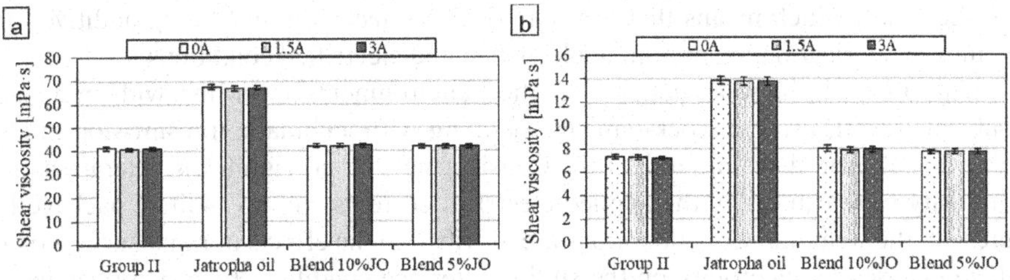

FIGURE 6.7 Mean shear viscosity of the different tested oils under non-electric (0 A) and electric (1.5 and 3 A) environments at: (a) 20°C; and (b) 75°C.

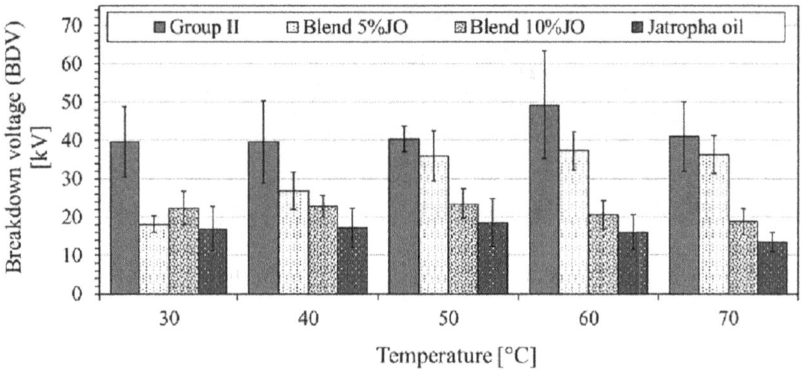

FIGURE 6.8 Dielectric strength or breakdown voltage (BDV) with temperature for the tested oils.

6.3.3 Dielectric Strength or Breakdown Voltage (BDV)

Figure 6.8 shows the BDV results obtained for the tested oils at different temperatures. Group II and Jatropha oil exhibited the highest (from 40 to 50 kV) and lowest (14 to 20 kV) BDV values, respectively, for all the tested temperatures. Regarding the blends, it can be seen that the addition of Jatropha oil to Group II generated a decrease of BDV at temperatures for all the cases. The low BDV of Jatropha oil can be ascribed to its possible higher moisture content. In general, VOs have higher water solubility levels than mineral oils, they can absorb 20–30 times more moisture at 25°C compared to mineral oil [20].

6.4 DISCUSSION

The lower wear and CoF generated by Jatropha oil for all the tested conditions in contrast to mineral oils can be mainly ascribed to its higher viscosity and its high content of saturated and monounsaturated fatty acids, such as oleic and hexadecanoic acids, which are known to improve the lubricity properties of VOs by forming more effective tribofilms at the sliding interfaces [7]. Both low additions of Jatropha oil into mineral oil were enough to promote a significant reduction of wear and CoF for all the cases, which means that it has a positive effect as a lubricating additive for boundary lubrication under both unelectrified and electrified conditions.

This increase in wear under electrified environments is in-line with previous published work using a block-on-ring operating with automatic transmission fluids used for electric driveline units [21]. Considering that no significant alterations in viscosity were caused by the applied electrical currents and no additive package is present, the wear and CoF alterations are ascribed to the effects of rapid surface oxidation caused by electricity on the sliding interface condition. Electricity has been demonstrated to generate an accelerated formation of iron oxides [4], but also carbon products that can act either as abrasive particles or lubricious films [22], which can generate accelerated wear by three-body abrasion mechanisms but also a possible reduction in CoF. Besides, these oxide and carbon products formed and retained at the surface asperities can lead to a decrease in film thickness [23] causing more wear in electrified lubricated steel sliding interfaces by scuffing mechanism, as shown in Figure 6.4b and c. In addition, the BDV of the oils is another influencing factor on the wear produced by electricity. Lubricants with high BDV may have poor conductivity but a significant ability to hold onto a charge with an applied electric field; the higher the BDV of the oil, the more powerful the electrical discharges when the dielectric strength of the lubricant film is broken, promoting more severe wear [4]. Hence, the low wear generated by Jatropha oil and the blends under electrified condition could be also associated with their lower BDV than mineral oil.

6.5 SUMMARY

The conducted electrified four-ball tests, electrorheological and BDV measurements were useful to find out more about the electrified tribological behavior of Jatropha oil as a potential bio-additive for electric drivelines lubricants. Jatropha oil was found to exhibit the lowest wear and CoF for both unelectrified and electrified tribological test conditions in comparison to mineral oil. Its addition in low concentrations (5 v/v% and 10% v/v) to a mineral oil caused a significant reduction in wear and CoF for both unelectrified and electrified tribological test conditions, which was associated with the low unsaturation degree of *Jatropha* oil, but also to its low BDV; this last for the electrified cases. The shear viscosity of all the oils did not change with the electrical currents applied, so no viscosity changes due to electrification were assumed through the electrified sliding tests. Finally, *Jatropha* oil was found to be a potential biodegradable additive to reduce wear and CoF not only in unelectrified contacts but also in electrified sliding contacts.

ACKNOWLEDGMENTS

C.C.R.H, C.D.R.C. and L.I.F.C. acknowledges the financial support given by Tecnologico de Monterrey within the Challenge-based research project IJST070-23EG68001 (call 2023) and the Fulbright-Comexus grant (Visiting Scholar-Postdoc 2023-2024).

REFERENCES

1. Chen, Y., Jha, S., Raut, A., Zhang, W., & Liang, H. (2020). Performance characteristics of lubricants in electric and hybrid vehicles: A review of current and future needs. *Frontiers in Mechanical Engineering*, 6. https://doi.org/10.3389/fmech.2020.571464.
2. Farfan Cabrera, L. I., & Erdemir, A. (2024). *Electric Vehicle Tribology: Challenges and Opportunities for a Sustainable Transportation Future*. Elsevier, Amsterm.
3. Farfan-Cabrera, L. I. (2019). Tribology of electric vehicles: A review of critical components, current state and future improvement trends. *Tribology International*, 138, 473–486. https://doi.org/10.1016/j.triboint.2019.06.029.
4. Farfan-Cabrera, L. I., Erdemir, A., Cao-Romero-Gallegos, J. A., Alam, I., & Lee, S. (2023). Electrification effects on dry and lubricated sliding wear of bearing steel interfaces. *Wear*, 516–517, 204592. https://doi.org/10.1016/j.wear.2022.204592.

5. Berman, D., Farfan-Cabrera, L. I., Rosenkranz, A., & Erdemir, A. (2024). 2D materials for durable and sustainable electric vehicles. *Nature Reviews Materials*. https://doi.org/10.1038/s41578-024-00680-3.
6. Zainal, N. A., Zulkifli, N. W. M., Gulzar, M., & Masjuki, H. H. (2018). A review on the chemistry, production, and technological potential of bio-based lubricants. *Renewable and Sustainable Energy Reviews*, 82, 80–102. https://doi.org/10.1016/j.rser.2017.09.004.
7. Kurre, S. K., & Yadav, J. (2023). A review on bio-based feedstock, synthesis, and chemical modification to enhance tribological properties of biolubricants. *Industrial Crops and Products*, 193, 116122. https://doi.org/10.1016/j.indcrop.2022.116122.
8. Mobarak, H. M., Niza Mohamad, E., Masjuki, H. H., Kalam, M. A., Al Mahmud, K. A. H., Habibullah, M., & Ashraful, A. M. (2014). The prospects of biolubricants as alternatives in automotive applications. *Renewable and Sustainable Energy Reviews*, 33, 34–43. https://doi.org/10.1016/j.rser.2014.01.062.
9. Farfan-Cabrera, L. I., Franco-Morgado, M., González-Sánchez, A., Pérez-González, J., & Marín-Santibáñez, B. M. (2022). Microalgae biomass as a new potential source of sustainable green lubricants. *Molecules*, 27(4), 1205. https://doi.org/10.3390/molecules27041205.
10. Panchal, T. M., Patel, A., Chauhan, D. D., Thomas, M., & Patel, J. V. (2017). A methodological review on bio-lubricants from vegetable oil-based resources. *Renewable and Sustainable Energy Reviews*, 70, 65–70. https://doi.org/10.1016/j.rser.2016.11.105.
11. Cermak, S. C., Isbell, T. A., Bredsguard, J. W., & Thompson, T. D. (2017). Estolides: Synthesis and applications. In: Ahmad, M. U. (ed.), *Fatty Acids: Chemistry, Synthesis, and Application*. Elsevier, Amsterdam, pp. 431–475.
12. Romsdahl, T., Shirani, A., Minto, R. E., Zhang, C., Cahoon, E. B., Chapman, K. D., & Berman, D. (2019). Nature-guided synthesis of advanced bio-lubricants. *Scientific Reports*, 9(1). https://doi.org/10.1038/s41598-019-48165-6.
13. Luna, F. M., Cavalcante, J. B., Silva, F. O., & Cavalcante, C. L. (2015). Studies on biodegradability of bio-based lubricants. *Tribology International*, 92, 301–306. https://doi.org/10.1016/j.triboint.2015.07.007.
14. Global Bio-Lubricants Market. (2024). Regulatory Norms Driving the Demand for Bio-Based Lubricants & Increasing Adoptionin the Automotive Industry. https://www.globenewswire.com/news-release/2019/12/23/1963974/0/en/Global-Bio-lubricants-Market-to-2024-Regulatory-Norms-Driving-the-Demand-for-Bio-based-Lubricants-Increasing-Adoption-in-the-Automotive-Industry.html (accessed on 10 August 2024).
15. Farfan-Cabrera, L. I., Gallardo-Hernández, E. A., Gómez-Guarneros, M., Pérez-González, J., & Godínez-Salcedo, J. G. (2020). Alteration of lubricity of jatropha oil used as bio-lubricant for engines due to thermal ageing. *Renewable Energy*, 149, 1197–1204. https://doi.org/10.1016/j.renene.2019.10.116.
16. Farfan-Cabrera, L. I., Gallardo-Hernández, E. A., Pérez-González, J., Marín-Santibáñez, B. M., & Lewis, R. (2019). Effects of jatropha lubricant thermo-oxidation on the tribological behaviour of engine cylinder liners as measured by a reciprocating friction test. *Wear*, 426–427, 910–918. https://doi.org/10.1016/j.wear.2019.02.028.
17. Farfan-Cabrera, L. I., Gallardo-Hernández, E. A., & Pérez-González, J. (2017). Compatibility study of common sealing elastomers with a biolubricant (jatropha oil). *Tribology International*, 116, 1–8. https://doi.org/10.1016/j.triboint.2017.06.039.
18. Aguilar-Rosas, O. A., Farfan-Cabrera, L. I., Erdemir, A., & Cao-Romero-Gallegos, J. A. (2023). Electrified four-ball testing: A potential alternative for assessing lubricants (E-fluids) for electric vehicles. *Wear*, 522, 204676. https://doi.org/10.1016/j.wear.2023.204676.

19. Cao-Romero-Gallegos, J. A., Reséndiz-Calderón, C. D., Cazarez-Ramírez, I., Galluzzi, R., & Farfan-Cabrera, L. I. (2023). Dielectric strength of electric vehicle fluids (lubricants and coolants) at usual operating temperatures. In: *2023 International Symposium on Electromobility (ISEM)*, Monterrey, Mexico. https://doi.org/10.1109/isem59023.2023.10334777.
20. Darwin, A., & Perrier, C. (2011). The use of natural ester fluids in transformer. In: *Proceedings MATPOST Conference,* vol. 36, pp. 15–16.
21. Lee, P. M., Sanchez, C., Frazier, C., Velasquez, A., & Kostan, T. (2023). Tribological evaluation of electric vehicle driveline lubricants in an electrified environment. *Frontiers in Mechanical Engineering*, 9. https://doi.org/10.3389/fmech.2023.1215352.
22. Deshpande, P., Yelkarasi, C., Lee, S., Farfan-Cabrera, L. I., & Erdemir, A. (2024). Electrotribochemical Formation of abrasive nano-carbon particles under electrified conditions on lubricated sliding contacts. *Carbon*, 228, 119425. https://doi.org/10.1016/j.carbon.2024.119425.
23. Cao-Romero-Gallegos, J. A., Taghizadeh, S., Aguilar-Rosas, O. A., Dwyer-Joyce, R. S., & Farfan-Cabrera, L. I. (2024). The effect of electrical current on lubricant film thickness in boundary and mixed lubrication contacts measured with ultrasound. *Friction*, 12(8), 1882–1896. https://doi.org/10.1007/s40544-024-0890-7.

7 Electrified Pin-on-Disk Testing for Evaluating Lubricated Contacts under Inert Gaseous Environments

Leonardo Israel Farfan Cabrera, Seungjoo Lee, Sean Skowron, and Ali Erdemir

7.1 INTRODUCTION

In many of the previous friction and wear studies, the effects of various lubricants and/or additives as well as test conditions have been investigated in detail using all kinds of tribomaterials and coatings. This includes varying parameters like lubricant viscosity, sliding speed, contact load, test temperature, roughness, contact geometry, test materials (including their mechanical properties), etc. However, the effects of electrical discharges (that occur in EV drivelines) on tribological performance has remained relatively unexplored [1]. The importance of studying the influence of electricity into the tribological behavior of bearings and gears from EV drivelines and the lubricity performance of driveline lubricants is because the operation of the electric motors in EVs can cause undesirable shaft voltages which can give rise to damaging bearing currents [2,3]. Bearing currents primarily result from magnetic flux asymmetry, inverter-induced voltage fluctuations, and triboelectricity, and they can collectively damage bearing interfaces [2,4], in particular, causing severe oxidation, lubricant degradation, micro-electro erosion/electrical discharge machining (EDM), cratering, fluting and micro/macro-pitting leading to catastrophic fatigue spalls, scoring, abrasive wear and scuffing in the tribological components, as demonstrated widely by different authors [5–12]. In addition to these problems, other recent studies have confirmed that the lubricant film thickness can be reduced significantly [13] either by changes in the viscosity and rheological behavior of the lubricant [14] and the products deposited in the surfaces due to electro-tribo-chemically induced reactions [15].

In an attempt to investigate the effect of electricity on the tribological performance of interacting surfaces and lubricants, a variety of test methodologies has been proposed in recent years. These included electrified bearing and gear tests for micropitting studies [16,17], tribological test machines with contact geometries of

 DOI: 10.1201/9781003496069-7

pin-on-disk [6,15,18,19], four balls [20], and block-on-ring [9]. Among others, due to its versatility and appropriateness for mimicking the boundary lubrication contacts of bearings and gears, pin-on-disk was one of the very first test methods used to explore the effects of electricity on the friction and wear of EV driveline materials under dry and lubricated sliding conditions [6,15,18]. In all these studies, it was confirmed that electricity was very detrimental to the overall performance of tribological test pairs. Specifically, much severe wear, scuffing, and micropitting were observed, and as a result, the test pairs failed much faster than those tested under unelectrified conditions. As for the underlying mechanisms, it was demonstrated that accelerated oxidation—leading to the formation of abrasive oxide particles, iron carbides, and soot-like carbonaceous wear debris—much more severe third-body abrasive wear and thermally induced softening of base steel, were all shown to occur and adversely affect the overall tribological performance of these steel test pairs [7,15].

Although significant advances (in terms of the understanding of electro-tribochemically induced reactions) have been achieved lately [15], further research on the particular contribution of different environmental species to these tribological problems is needed to understand the complex interactions in an electrified, lubricated sliding interface. Therefore, other, more sophisticated test methods are required. This chapter aims to propose a pin-on-disk method to systematically investigate the effect of electricity and controlled inert-gas atmospheres on friction and wear of lubricated interfaces. The idea of the proposed test set-up and protocol is to enable evaluations of wear and friction of lubricated metallic interfaces, eradicating the influence of tribo-oxidation, which has been shown to have a significant, detrimental influence on the tribological properties under electrification. To illustrate the method, pin and disk test pairs made of AISI 52100 steel were tested under unelectrified and electrified lubricated sliding conditions in ambient air and inert gas (N_2) atmospheres, respectively. Therefore, in this chapter, we present a comparison of wear, coefficient of friction (CoF), and electrical contact resistance (ECR) obtained on steel test pairs in ambient air and N_2 atmospheres.

7.2 MATERIALS AND METHOD

The tests were performed using steel ball and disk pairs in a tester (Anton Paar, TRB3), equipped with an air-tight plexiglass enclosure for maintaining a controlled inert gas atmosphere, as shown in the sketch in Figure 7.1. This method has been used and reported in detail elsewhere [21]. The tester was modified to carry out unelectrified and electrified tests, implementing a two-electrode cell system for applying a controlled external DC current to the sliding contact interface. To prove the presented method, three different low-viscosity commercial lubricants were tested under the conditions shown in Table 7.1. The specification and viscosity of the tested lubricants can be seen in Table 7.2.

To allow proper insulation of the lubricated contact, special ball and disk holders were machined out of polyether ether ketone (PEEK) with a very high dielectric constant [22]. A DC power supply device (Keithley, 2230-30-1) was used as an external source to apply electricity to the contact. It is noteworthy that either a DC or AC under other magnitudes of current/voltage can also be connected; it depends on each

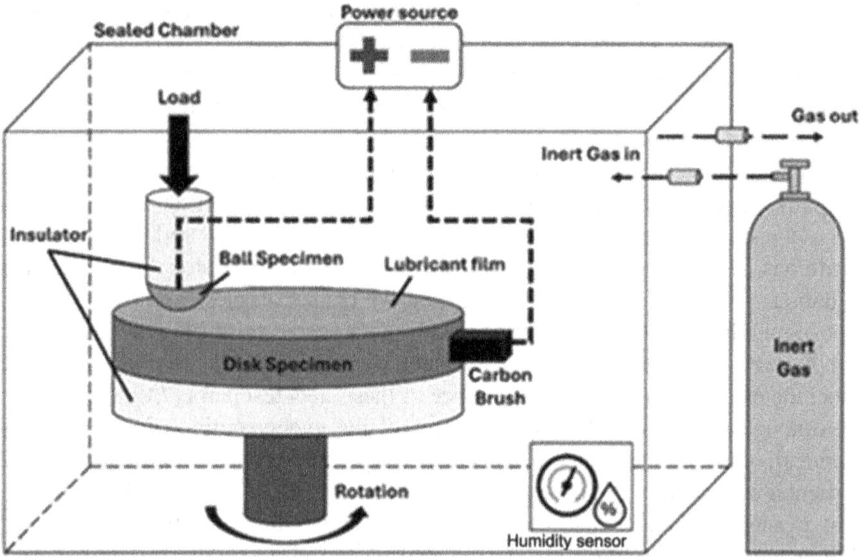

FIGURE 7.1 Schematic representation of the pin-on-disk test set-up for the electrified tests under inert-gas atmospheres.

TABLE 7.1
Test Conditions

Parameter	Values
Atmosphere	Ambient air; N_2
Load (N)	5
Contact pressure (MPa)	700
Oil quantity per test (μL)	10
Speed (m/s)	0.1
Temperature (°C)	22.5 ± 1
Sliding distance (m)	1000
DC current (A)	0, 3
Test repeats	3 for ambient air; 2 for N_2
Materials	Balls: AISI 52100 steel (12.7 mm diameter), 0.02 μm surface roughness, Sa. Disks: AISI 52100 steel (80 mm in diameter), 64 HRC and 0.02 μm surface roughness, Sa.

electrical condition to be replicated. The DC power supply was connected to the ball and to the disk. For the disk, a carbon brush was loaded against the disk to maintain an electrified contact during disk rotation. The magnitude of the applied current (3 A) was selected to generate significant wear differences under the contact conditions tested, thus saving N_2 gas consumption as much as possible. In addition, a

TABLE 7.2
Specification and Kinematic Viscosity of the Lubricants Tested

Fluid Property		ATF (Synthetic Base ATF Oil, Mobil)	DEXRON (Mineral-Base Dexron VI, Mobil)	Gear Oil (Synthetic Base Oil, SAE 75W-140, Valvoline)
Kinematic viscosity	40°C	36.3	29.5	181
(cSt)	100°C	7.4	5.8	27.7

voltage data logger was connected in parallel to measure and log the voltage, V, with sliding distance at the contact interface when a constant DC current, I, was applied. It was used then to calculate the ECR by Ohm's law equation:

$$\text{ECR} = \frac{V}{I} \tag{7.1}$$

ECR data is useful to have an insight into the contact resistivity variation caused by alteration into the film thickness, wear, and possible tribolayer formation (i.e., oxides have much higher resistivity than steel on steel contact) [20].

For all these tests, a small amount of oil (10 μL) was spread over the sliding track on the disk sample, which was large enough to maintain the wear track lubricated during the entire test and to prevent oil from entering the carbon brush interface. The CoF was measured and logged by the tribometer's data acquisition software for post-test analyses. Both unelectrified and electrified tests were run in ambient air and in an inert gas atmosphere (for which the chamber was continuously fed with N_2 gas through a supply inlet in order to create and maintain a positive pressure inside the test chamber, to prevent air from backfilling the chamber, and to reduce relative humidity below 3%). For the test protocol demonstration, three test repeats were carried out for the tests under air atmosphere, while two test repeats were done for the N_2 atmosphere cases.

After the tests, the wear volume of the balls was measured using an optical profilometer (Countour GT, Bruker) while the wear track of the disks was estimated by Equation (7.2) according to the ASTM G-99 standard method.

$$\text{Disk volume loss} = \frac{\pi(r)(w)^3}{6R} \tag{7.2}$$

where r is the track radius, w is the track width, and R corresponds to the ball radius. The track width was measured by the optical profilometer. Four wear track profile measurements were taken with a 90° offset from each track.

7.4 RESULTS AND DISCUSSION

The comparisons of ball and disk wear volumes obtained for all the tests are shown, respectively, in Figure 7.2a and b (data obtained from Ref. [21]). Firstly, comparing the results obtained under ambient air, it is evident that the three lubricants presented

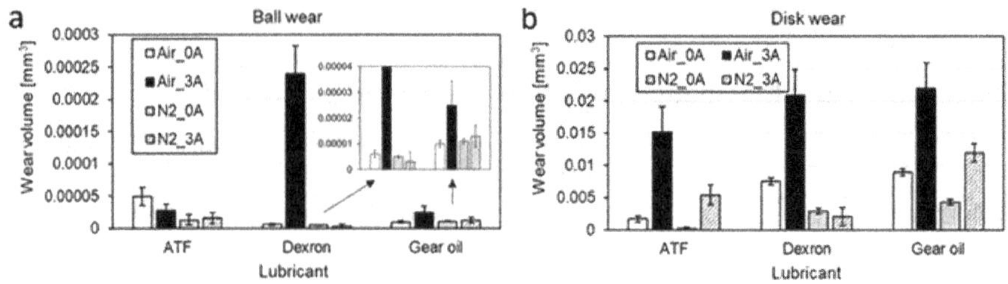

FIGURE 7.2 Comparison of wear volumes obtained by unelectrified and electrified testing under air and N_2 atmospheres: (a) ball wear and (b) disk wear.

an alteration of wear for both ball and disk samples due to electrification. The electrical condition promoted a significant increase (>100%) in ball and disk wear for Dexron and gear oil. In the case of ATF, electricity produced a decrease (about 40%) in the ball wear but a much higher increase (about 750%) in disk wear. This is because there was a build-up in the ball scar but severe abrasion in the disk surface. Therefore, in general, the applied electricity generated very dramatic wear increases using the three oils under ambient air, which is associated with accelerated oxidation and formation of abrasive nano-carbon particles that collectively trigger more severe abrasion and scuffing, as reported previously [7,15].

Comparing the results for the three lubricants tested in ambient air and N_2 atmospheres under the unelectrified condition, N_2 atmosphere was able to promote a significant reduction of wear (>50%), particularly in the disk samples of the three lubricants. In previous research works, it has been ascribed to the suppression of tribo-oxidation by the N_2 atmosphere, which in turn significantly reduces the contribution of oxide particle abrasion and facilitates the formation of nitrides and/or beneficial carbon boundary films [23,24]. In a similar way, but comparing the results for the three lubricants tested under electrified conditions, the N_2 atmosphere was also found to be even far more beneficial to decreasing wear in both ball and disk samples. This significant wear decrease obtained for the three lubricants is associated with the effective suppression of electro-tribo-oxidation under the inert-gas atmosphere. In order to illustrate the wear produced on the samples by the test method proposed, Figure 7.3a–d show comparisons of the wear scars produced on the balls and disks tested using ATF under unelectrified and electrified conditions in ambient air and N_2 atmospheres, respectively. In all the cases, the significant reduction of wear in the disk tracks by N_2 atmosphere was evident.

Within this test method, it was possible to measure CoF and ECR (this only for the electrified cases) in real time. Figure 7.4a–c show the comparison of CoF (left) and ECR (right) behaviors for the three lubricants in ambient air and N_2 atmospheres. In the case of the CoF behaviors obtained in ambient air, electricity promoted an apparent increase of CoF for the three oils. Also, the CoF behavior obtained under the electrified condition seems to be more unstable (with larger variations in amplitude), which is related to the accelerated surface oxidation and more severe adhesion and abrasion situations caused by electricity. Besides, it is important to note that CoF started being higher for the three oils under electrified conditions from the beginning of the test (from 20 to 100 m sliding distance). It is related to the possible reduction

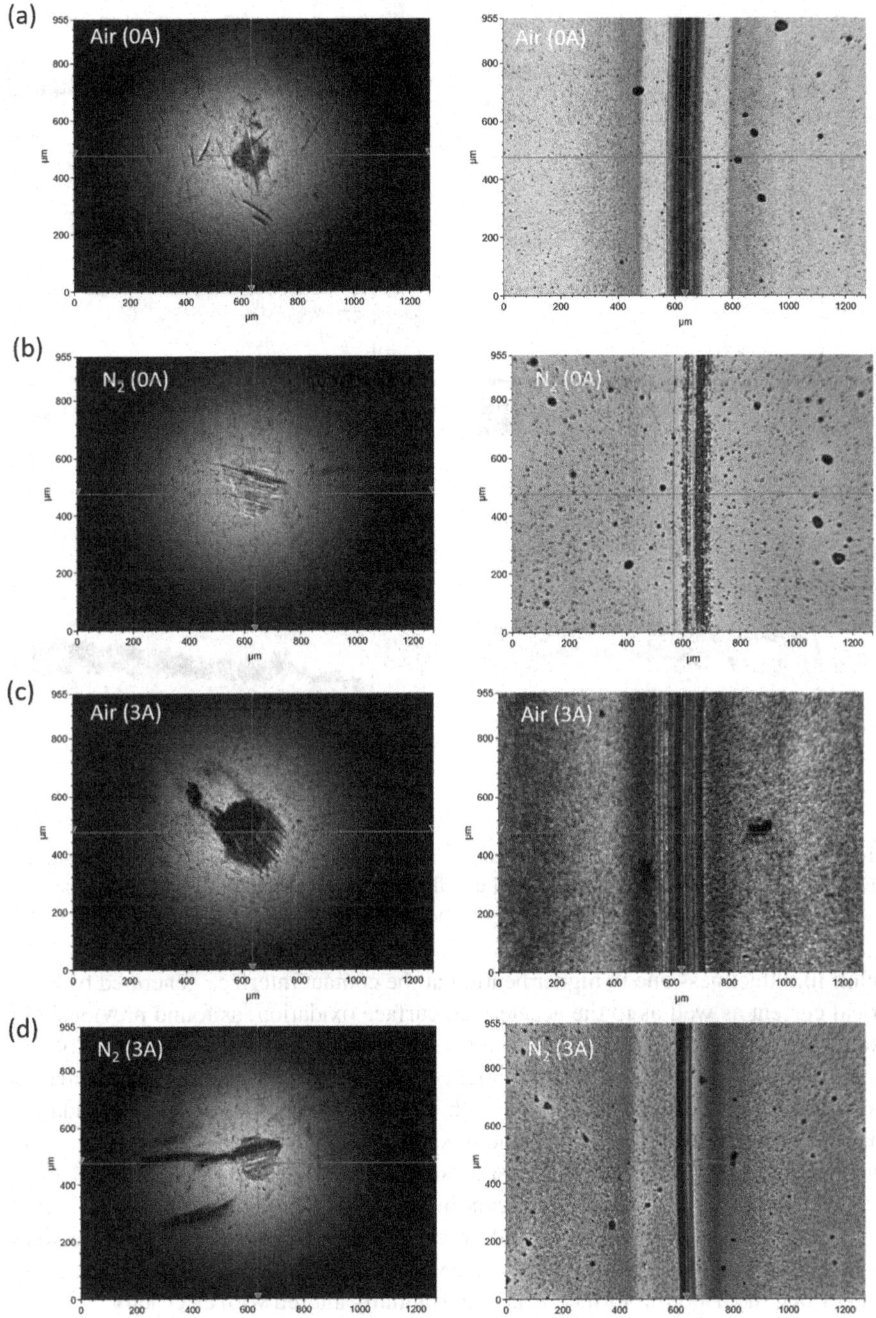

FIGURE 7.3 Comparison of wear scars formed on ball and disk samples when using ATF under the different conditions: (a) ambient air (0 A); N_2 atmosphere (0 A); ambient air (3 A); and N_2 atmosphere (3 A).

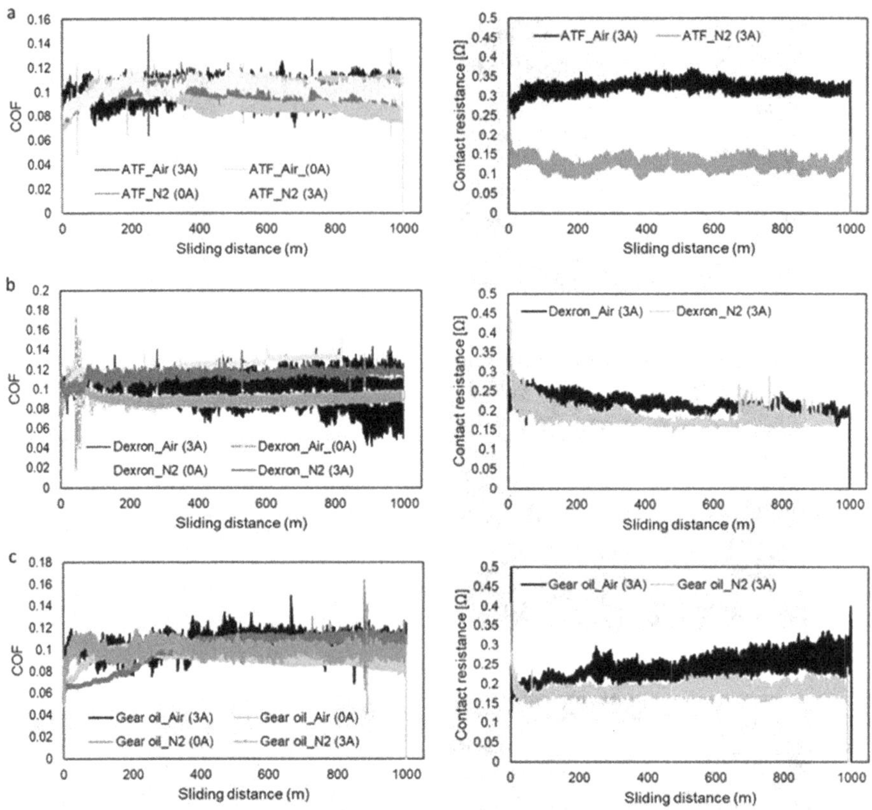

FIGURE 7.4 Comparisons of coefficient of friction (CoF) and electrical contact resistance (ECR) under unelectrified and electrified conditions in ambient air and N_2 atmospheres for the different tested oils: (a) ATF; (b) dexron; and (c) gear oil.

of oil film thickness due to higher heating at the contact interface generated by electrical current as well as to the accelerated surface oxidation, as found previously in Ref. [13]. On the other hand, the N_2 atmosphere generated slightly higher CoFs in comparison to those obtained in ambient air for both unelectrified and electrified conditions, which can be associated with that substantial suppression of oxidation since it is well known that oxides tend to serve as solid lubricant, to some extent, in metallic interfaces. However, the most outstanding finding with the N_2 atmosphere is that electricity did not have a significant influence on the CoF behavior in this inert gas atmosphere, neither the amplitude nor magnitude, which means that the wear mechanisms that occurred at the interface do not change too much with electricity and/or that the tribofilm formed is not substantially altered with electricity.

The ECR results obtained by this test were helpful for elucidating the above. The ECR obtained for the three oils in ambient air tended to be higher than that obtained in N_2 atmosphere. However, at the beginning of the tests (0–20 m sliding distance), ECR magnitude was similar for both ambient air and N_2 atmospheres, then it tended to increase with sliding distance under ambient air but decreased under N_2

atmosphere for the three oils. If an ideal ball on flat contact (where no tribofilms nor wear products are formed) is considered, wear would increase the real contact area, so ECR would be expected to decrease with sliding distance. Nonetheless, the ECR data obtained by these experiments show that ECR reduction with sliding distance only happened under N_2 atmosphere; then it tended to increase with sliding distance under ambient air, which could only be ascribed to the formation of a tribofilm or wear products with higher resistivity, i.e., oxides. As a reference, 52100 steel has a resistivity in the order of 10^{-4} Ω cm, but iron oxide (hematite single crystals) has a resistivity in the order of 10^{6} Ω cm [25]. It is noteworthy that the reduction of ECR with sliding distance generated under N_2 atmosphere can be attributed to the increase in contact area by wear, but also to a possible formation of other low resistivity tribofilms, i.e., iron nitride films (also in the order of 10^{-4} Ω cm [26]) and carbonaceous products (i.e., black carbon resistivity is in the order of 10^{-2} Ω cm [27]).

7.5 SUMMARY

The present pin-on-disk set-up and method were shown to be very useful to test the different lubricated metallic contacts under both unelectrified and electrified conditions in ambient air and inert gas (N_2) atmospheres, generating measurable wear volumes with significant differences and obtaining CoF and ECR data through the entire tests. The test was versatile in terms of the materials and geometries to be tested as well as the electrical current type and magnitude to be replicated. In these experiments, N_2 gas was used; however, any other inert gas can also be utilized. Getting CoF and ECR data simultaneously in any type of atmosphere is valuable to detect the possible chemical or phase changes at the interface (i.e., tribofilm and/or wear) with the sliding distance under electrified conditions. It is important to mention that a very low humidity (<3%) achieved by the amount of N_2 gas was found to be effective in generating a significant and consistent reduction in wear. It was observed that higher humidity magnitudes tended to promote not repeatable results or similar wear and CoF values than those obtained in ambient air. Hence, the control of humidity in the test by applying the proper amount of inert gas was found to play an important role in the consistency of the results by this method.

According to the findings obtained by these tests, the application of N_2 atmosphere allowed a significant reduction of wear for the three oils tested under both unelectrified and electrified conditions in comparison with the tests under ambient air. The CoF behaviors generated under the N_2 atmosphere in both unelectrified and electrified conditions tended to be more stable (with lower amplitudes) than those generated under ambient air, in particular under electrified conditions. Besides, the obtained ECR with sliding distance under electrified conditions depicted significant differences between tests in ambient air and N2 atmosphere. For the three lubricants tested, ambient air promoted ECR to increase with sliding distance; meanwhile, N_2 promoted ECR to decrease with sliding distance. Both ECR trends were ascribed to different tribofilms (with different resistivity) formed possibly at the interfaces. Overall, the highly instrumented ball on-disk machine (see Figure 7.1) developed by our team was very useful in studying the friction and wear of test pairs under electrified and unelectrified conditions and in ambient air and inert gas environments.

REFERENCES

1. Seungjoo, L., Cao-Romero-Gallegos, J. A., Farfan-Cabrera, L. I., & Erdemir, A. (2024). Tribological challenges for a sustainable e-mobility, in: Farfan-Cabrera, L. I., & Erdemir, A. (Eds.), *Series on Tribology and Surface Engineering, Electric Vehicle Tribology*, Elsevier, pp. 5–21. https://doi.org/10.1016/B978-0-443-14074-7.00002-9.
2. He, F., Xie, G., & Luo, J. (2020). Electrical bearing failures in electric vehicles. *Friction*, 8(1), 4–28. https://doi.org/10.1007/s40544-019-0356-5.
3. Farfan-Cabrera, L. I. (2019). Tribology of electric vehicles: A review of critical components, current state and future improvement trends. *Tribology International*, 138, 473–486. https://doi.org/10.1016/j.triboint.2019.06.029.
4. Tawfiq, K. B., Güleç, M., & Sergeant, P. (2023). Bearing current and shaft voltage in electrical machines: A comprehensive research review. *Machines*, 11(5), 550. https://doi.org/10.3390/machines11050550.
5. Xie, G., Guo, D., & Luo, J. (2015). Lubrication under charged conditions. *Tribology International*, 84, 22–35. https://doi.org/10.1016/j.triboint.2014.11.018.
6. Cao-Romero-Gallegos, J. A., Farfan-Cabrera, L. I., Erdemir, A., & Pascual-Francisco, J. B. (2023). Lubricated sliding wear of gear material under electrification: A new approach to understanding of the influence of shaft currents in the wear of EV transmissions. *Wear*, 523, 204782. https://doi.org/10.1016/j.wear.2023.204782.
7. Farfan-Cabrera, L. I., Erdemir, A., Cao-Romero-Gallegos, J. A., Alam, I., & Lee, S. (2023). Electrification effects on dry and lubricated sliding wear of bearing steel interfaces. *Wear*, 516–517, 204592. https://doi.org/10.1016/j.wear.2022.204592.
8. Schneider, V., Krewer, M., Poll, G., & Marian, M. (2024). Effect of harmful bearing currents on the service life of rolling bearings: From experimental investigations to a predictive model. *Lubricants*, 12(7), 230. https://doi.org/10.3390/lubricants12070230.
9. Lee, P. M., Sanchez, C., Frazier, C., Velasquez, A., & Kostan, T. (2023). Tribological evaluation of electric vehicle driveline lubricants in an electrified environment. *Frontiers in Mechanical Engineering*, 9. https://doi.org/10.3389/fmech.2023.1215352.
10. Schneider, V., Behrendt, C., Höltje, P., Cornel, D., Becker-Dombrowsky, F. M., Puchtler, S., Gutiérrez Guzmán, F., Ponick, B., Jacobs, G., & Kirchner, E. (2022). Electrical bearing damage, a problem in the nano- and macro-range. *Lubricants*, 10(8), 194. https://doi.org/10.3390/lubricants10080194.
11. Gould, B., Demas, N., Erck, R., Lorenzo-Martin, M. C., Ajayi, O., & Greco, A. (2021). The effect of electrical current on premature failures and microstructural degradation in bearing steel. *International Journal of Fatigue*, 145, 106078. https://doi.org/10.1016/j.ijfatigue.2020.106078.
12. Morales-Espejel, G. E. (2024). Damaging modes of rolling bearings in electrical vehicle applications—Part 1, in: Farfan-Cabrera, L. I., & Erdemir, A. (Eds.), *Series on Tribology and Surface Engineering, Electric Vehicle Tribology*, Elsevier, pp. 107–130. https://doi.org/10.1016/B978-0-443-14074-7.00007-8.
13. Cao-Romero-Gallegos, J. A., Taghizadeh, S., Aguilar-Rosas, O. A., Dwyer-Joyce, R. S., & Farfan-Cabrera, L. I. (2024). The effect of electrical current on lubricant film thickness in boundary and mixed lubrication contacts measured with ultrasound. *Friction*, 12(8), 1882–1896. https://doi.org/10.1007/s40544-024-0890-7.
14. Lee, P., Jackson, M., & Sanchez, C. (2024). Electric vehicle fluid testing, in: Farfan-Cabrera, L. I., & Erdemir, A. (Eds.), *Series on Tribology and Surface Engineering, Electric Vehicle Tribology*, Elsevier, pp. 277–290. https://doi.org/10.1016/B978-0-443-14074-7.00016-9.
15. Deshpande, P., Yelkarasi, C., Lee, S., Farfan-Cabrera, L. I., & Erdemir, A. (2024). Electrotribochemical formation of abrasive nano-carbon particles under electrified conditions on lubricated sliding contacts. *Carbon*, 228, 119425. https://doi.org/10.1016/j.carbon.2024.119425.

16. Graf, S., Werner, M., Koch, O., Götz, S., & Sauer, B. (2023). Breakdown voltages in thrust bearings: Behavior and measurement. *Tribology Transactions*, 66(3), 488–496. https://doi.org/10.1080/10402004.2023.2185560.
17. Loos, J., Bergmann, I., & Goss, M. (2021). Influence of high electrical currents on WEC formation in rolling bearings. *Tribology Transactions*, 64(4), 708–720. https://doi.org/10.1080/10402004.2021.1909789.
18. Farfan-Cabrera, L. I., Cao-Romero-Gallegos, J. A., Lee, S., Komurlu, M. U., & Erdemir, A. (2023). Tribological behavior of H-DLC and H-free DLC coatings on bearing materials under the influence of DC electric current discharges. *Wear*, 522, 204709. https://doi.org/10.1016/j.wear.2023.204709.
19. Crilly, L., Jackson, R. L., Bond, S., Mills, G., & Bhargava, S. (2020). An investigation of the electrical contact resistance change, lubrication, and wear properties of a nanolubricant, in: *2020 IEEE 66th Holm Conference on Electrical Contacts and Intensive Course (HLM)*, San Antonio, TX. https://doi.org/10.1109/hlm49214.2020.9307917.
20. Aguilar-Rosas, O. A., Farfan-Cabrera, L. I., Erdemir, A., & Cao-Romero-Gallegos, J. A. (2023). Electrified four-ball testing – a potential alternative for assessing lubricants (E-fluids) for electric vehicles. *Wear*, 522, 204676. https://doi.org/10.1016/j.wear.2023.204676.
21. Farfan-Cabrera, L. I., Lee, S., Skowron, S., & Erdemir, A. (2024). Enhancing lubrication of electrified interfaces by inert gas atmosphere. *Journal of Tribology*, 147(5). https://doi.org/10.1115/1.4066649.
22. Niu, S., Bai, X., Cheng, Y., Qin, H., Zhang, Y., & Jiang, Z. (2023). Tuning high-temperature dielectric properties of poly ether ketone by using self-crosslinkable polyetherimide and nanoparticles. *Polymer Testing*, 117, 107858.
23. Zhang, J., Campen, S., Wong, J., & Spikes, H. (2022). Oxidational wear in lubricated contacts – or is it? *Tribology International*, 165, 107287. https://doi.org/10.1016/j.triboint.2021.107287.
24. Erdemir, A., Ramirez, G., Eryilmaz, O. L., Narayanan, B., Liao, Y., Kamath, G., & Sankaranarayanan, S. K. (2016). Carbon-based tribofilms from lubricating oils. *Nature*, 536(7614), 67–71. https://doi.org/10.1038/nature18948.
25. Lindgren, T., Vayssieres, L., Wang, H., & Lindquist, S. E. (2003). Photo-oxidation of water at hematite electrodes, in: Kokorin, A. I., & Bahnemann, D. (Eds.), *Chemical Physics of Nanostructured Semiconductors,* VSP, Utrecht, Boston, pp. 83–110.
26. Naganuma, H., Nakatani, R., Endo, Y., Kawamura, Y., & Yamamoto, M. (2004). Magnetic and electrical properties of iron nitride films containing both amorphous matrices and nanocrystalline grains. *Science and Technology of Advanced Materials*, 5(1–2), 101–106. https://doi.org/10.1016/j.stam.2003.10.020.
27. Spahr, M. E., Gilardi, R., & Bonacchi, D. (2013). Carbon Black for electrically conductive polymer applications, in: Mohanty, M., Misra, M., & Drzal, L. T. (Eds.), *Encyclopedia of Polymers and Composites*, Springer, pp. 1–20. https://doi.org/10.1007/978-3-642-37179-0_32-1.

8 Tribo-Impedance Characterization of EV Fluids

Yan Chen, Mohsen Tajedini, and Hong Liang

8.1 INTRODUCTION

Typically, tribological tests for lubricants need to simulate the working condition of contacting moving parts. When combined with physical property tests such as rheological and electrical resistance tests, the properties of a working lubricant can be measured. One of the most common configurations of tribotesting is pin-on-disc or ball-on-disc. This method is used widely for research and study. It is carried out using a pin or bearing ball that pushes against a revolving disc while recording the friction force. Tribotesting with such a configuration can be used to evaluate tribological characteristics such as anti-friction and -wear performance or lubricity.[1,2]

Previous studies on probing the electrical properties of materials during sliding contact have focused on the measurement of contact resistance[3–4] or the nature of an oil film.[5–6] Experiments have been conducted by probing the direct current conductivity or alternative current conductivity of a tribopair. The contact resistance was obtained through the DC current resistance, while the film thickness was obtained by analyzing the AC current resistance.[7,8] In order to simultaneously measure the thickness of a lubricating film and the electric resistance, we used an approach that superimposes the impedance measurement on friction monitoring.

8.2 SUMMARY OF METHODS TO EVALUATE THE PERFORMANCE OF LUBRICANTS

Lubricated tests that trigger electric damage can be performed on the motor or under lab-controlled conditions. To test the performance of a lubricant in real-life conditions, motor bearings powered with inverter can be tested in simulated test rigs. The bearing current and wear can be directly measured[9] with a simulated setup. In those experiments, a replicated voltage from an inverter was applied through the bearing to simulate the working conditions inside electric vehicles (EV) and hybrid electric vehicles (HEV). The experiment can also be performed on the actual motors inside an EV/HEV. By using the inductive measurement on the electric motors, the so-called hardware-in-the-Loop (HiL) setup provides more accurate measurement and can be integrated into EV/HEV.

 DOI: 10.1201/9781003496069-8

The most common experimental method for measuring the thermal conductivity of a lubricant is called the transient hot-wire method.[10] The transient hot-wire experimental setup is simple to perform and has high accuracy. This method uses a Pt or Ni wire, which is sealed inside a cylindrical pressure vessel filled with lubricant. The wire is heated for a short amount of time electrically, and its temperature is monitored simultaneously by its electric resistance. The thermal conductivity and the thermal capacity of the lubricant can be calculated from the temperature change of the wire. Essentially, this measurement setup can be modeled as an axisymmetric thermal transportation problem. It has an additional advantage when used to characterize lubricants, as the lubricant's thermal properties are highly correlated with its pressure, and the pressurized transient hot-wire method is relatively easy to achieve.

For small quantity lubricant measurement, a laser flash method can be used for measuring the thermal diffusivity.[11] This system uses a laser beam to heat the lubricant and optically measure the temperature change. Instead of an axisymmetric rod, this method models the system as an infinite-sized slab. The laser heats an infinitely thin layer of lubricant, and the temperature change thus can be fitted with a function of thermal diffusivity and thermal capacity. This method has an advantage when used for small batch experiments as it only requires a minimal amount of lubricant. Testing a lubricant's thermal properties requires a controlled heat source and an accurate temperature monitoring system. The system's accuracy and precision depend on a simple and easy-to-model measurement setup. Both the transient hot-wire method and laser flash method use thermal transportation equations with reduced dimensions. The laser flash method has the advantage of testing a small quantity of lubricant.

8.3 IMPEDANCE MEASUREMENT FOR EV LUBRICANTS

The electrochemical impedance can be measured through the response of electrodes to minor changes in potentials at various frequencies. These parameters defined in impedance are meaningful to the kinetics of changing surfaces. It is thus beneficial to study electrified surfaces. The $I(t)$ is the current, $V(t)$ the potential, w is an angular frequency, and $Z(\omega)$ the impedance. Here,

$$Z(\omega) = V(t) / I(t) \tag{8.1}$$

Here t is time.

$$V(t) = V0 \sin \alpha x \tag{8.2}$$

$$I(t) = I0 \sin(\alpha x + \theta) \tag{8.3}$$

Here θ = phase angle between $V(t)$ and $I(t)$.

During an electrochemical process, when the surface energy changes, it induces a time lag with a measurable phase angle, θ. Through simulation, the impedance can be expressed by real and imaginary parts. According to those, a tribo-electro system was built that measures the impedance of a lubricant while the friction is monitored. The configuration of this system is illustrated in Figure 8.1. The system

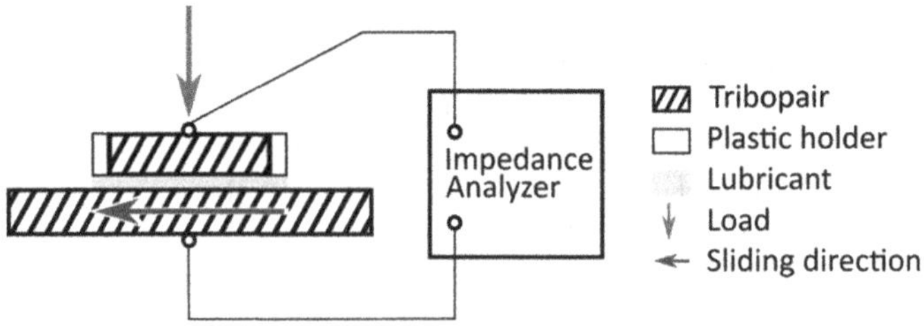

FIGURE 8.1 The configuration of a tribo-electro measurement system.[18]

consists of a disc-on-disc metallic pair connected to an instrument with impedance measuring capability. The upper circular disc is stationary. A normal force is applied on the disc, while the lower disc rotates. The contact area is known. Impedance was measured at a frequency when the friction becomes stable, and a constant voltage (V) was preset with a DC bias. This system makes it possible for *in situ* monitoring of the thickness and electrical resistance of lubricants simultaneously.

The thickness and resistance of a lubricant can be obtained through calculations based on impedance data. Here, the tribopair is considered as a capacitor and a resistor in parallel. The capacitance term is assumed to be fully contributed by the oil film, which has a specific dielectric constant [207].[12] Here are the formulas for calculation:

$$R = \frac{1}{\mathrm{Re}\left(\frac{1}{Z}\right)} \tag{8.4}$$

$$t = A\frac{\varepsilon\varepsilon_r}{\mathrm{Im}\left(\frac{1}{Z}\right)/\omega} \tag{8.5}$$

where R is the resistance, Z is the measured impedance subtracted from the impedance of the shorted measuring system $Z = Z_{\text{measured}} - Z_{\text{shorted}}$, A is the nominal area of contact, $\varepsilon\varepsilon_r$ is the dielectric constant, ω is the angular frequency of the applied voltage. Re and Im takes the real and imaginary part of a complex number, respectively.

If an AC current is applied to the contact interface, the voltage will fluctuate and can be recorded, as illustrated in Figure 8.2. During a tribotest, the impedance can be measured synchronically.[13] The schematic of an equivalent electrical circuit represents a lubricated contact under the influence of an applied AC current. This model conceptualizes the lubricated interface as a combination of variable capacitive and resistive elements in parallel.[14] Variable components were chosen because the impedance of contact changes with the variation of contact potential, film thickness, and tribochemical reactions. The capacitance primarily stems from the dielectric properties of the lubricant, whereas the resistance is attributed to the physical contact between asperities, oxide layers, and tribofilm that may form during testing.[15] By distinguishing between the capacitance and resistance of the interface, lubricant formulations can be optimized for specific electrified conditions.

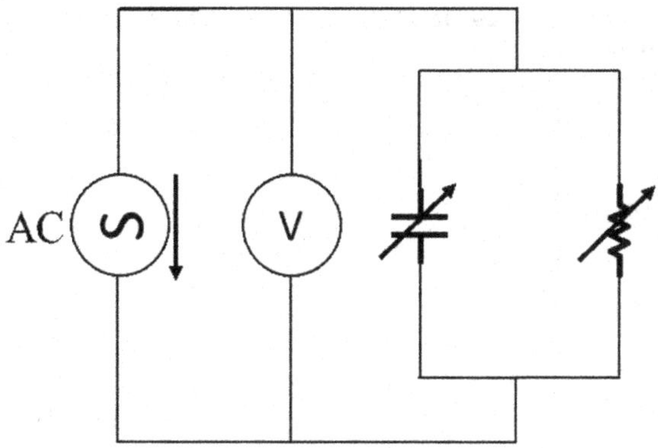

FIGURE 8.2 An equivalent circuit of a lubricated contact with an applied AC current. Figure obtained from one of our recent studies.[20] (Figure Use with permission.)

The tribological conditions for this impedance evaluation are kept the same as the wear test, while an impedance analyzer can be employed rather than the DC power supply. To monitor impedance changes, in a recent study on grease for EVs, the frequency was set at 20 kHz, the DC-biased voltage at 0.5 mV, and the AC voltage at 1 mV rms. The choice of frequency is similar to the EV's motor conditions.[16] Low DC-biased voltage and AC voltage were chosen to minimize possible discharges and focus on the impedance variations. DC-biased voltage was used to assure that discharge damage would happen on the sample disc.

The total impedance of the contact $\left(Z_{(t)}\right)$ can be calculated as:

$$\frac{1}{Z_{(t)}} = \frac{1}{Z_R} + \frac{1}{Z_c} \tag{8.6}$$

where Z_R is the impedance of the resistor and Z_c denotes the impedance of the capacitor in the circuit, they can be respectively expressed as follows:

$$Z_R = R \tag{8.7}$$

$$Z_c = \frac{1}{j\omega C} \tag{8.8}$$

The variables ω, R, and C represent the angular frequency, contact resistance, and the capacitance of the contact, respectively. The total impedance can be calculated as follows to be able to distinguish the real and imaginary parts:

$$Z_{(t)} = \frac{R}{1+\omega^2C^2R^2} - j\frac{\omega CR^2}{1+\omega^2C^2R^2} \tag{8.9}$$

The equation correlates the electrical inputs to impedance and wear protection in electrified systems. If the imaginary part of a contact is higher, the chance of discharges is higher. As the real contact area and film thickness can change at different speeds, the impedance of the contact will change. The impedance can be measured while increasing the sliding speed. The impedance of contact consists of capacitance

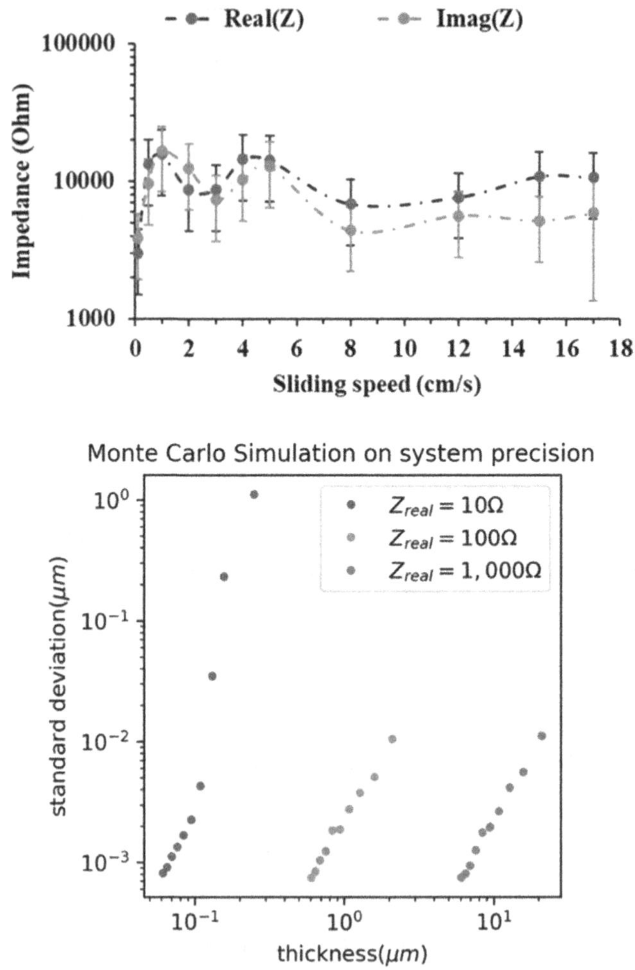

FIGURE 8.3 (a) An example of impedance measured against sliding speed. Reported in one of our recent studies.[20] (Figure use with permission). (b) The Monte Carlo simulation on system precision.

and resistance. When the impedance of the contact is more capacitive, frequent discharge events are likely as they break the lubricant film. Conversely, a more resistive behavior indicates that the current predominantly flows through the asperity contacts, maintaining a steadier electrical flow through the entire contact. An example of impedance measurement against the sliding speed is shown in Figure 8.3.

8.4 LUBRICANT FILM THICKNESS

In order to obtain properties of the lubricating oil, the impedance measured previously was superimposed on a Stribeck curve, i.e., the friction against a grouping of (speed × viscosity) / load, as the Sommerfeld grouping.[17] The precision of this measuring system can be inferred from the shorted system impedance. The standard deviation of the shorted measurement system was ~0.05Ω for both the real and imaginary part of the impedance. The precision of film thickness measurement was dependent on the real part of measured impedance and film thickness. The analytical solution is difficult, but it can be estimated by the Monte Carlo simulation method, see Figure 8.3. The Z_{real} = 10, 100, 1000 Ω was simulated with a total 200 simulated experiments:

The range of this measuring system was limited by the current measuring capability of our instruments. The high impedance can cause the current to be too low to detect. This instrument can measure 60 pA current; thus, it can detect maximum impedance of 16 MΩ.

8.5 THERMAL BEHAVIOR

The thermal behavior of a lubricating film can be probed using the same setup previously discussed. The system measures the thickness of a lubricating film while imposing a thermal gradient into it. This thermal gradient is monitored with a thermocouple and correlated with the oil film thickness. A model to relate the measured quantity to the thermal conductivity of the lubricating film can be calculated. Temperature is controllable during experiments through a heating element attached underneath the lower disc. During testing, the temperature of the lubricant is evenly increased. This temperature increment can be monitored using a thermocouple attached to the upper disc. The position of this thermocouple is recommended to be about 0.5 mm above the edge of the upper disc. In experiments, the temperature range can be preset pending the conditions the heating blocks provide. In the case here, it was controlled from room temperature to around 320 K. Here, the temperature range was limited by the plastic materials used to construct this system. At a fixed temperature, the impedance of this system remains constant. Figure 8.4 shows the directly measured the impedance at room temperature. This shows impedance value was unchanging when the temperature was fixed. Without temperature change, the electric response of the contacting surface is relatively stable.

8.6 CONCLUSIONS

The challenging conditions in electric vehicles demand identification and understanding of fundamental properties of fluids. Methods to measure their properties, particularly during operation, need to be further developed. New approaches, as well as specifications, are needed to meet the new demands of EVs.

The impedance method reported here is effective in revealing some basic properties and discovering characteristic performance of EV fluids. It shows the high potential of this method for advancing EV technology and future fundamental investigation. Continued effort using this method will enable the design of better EV fluids.

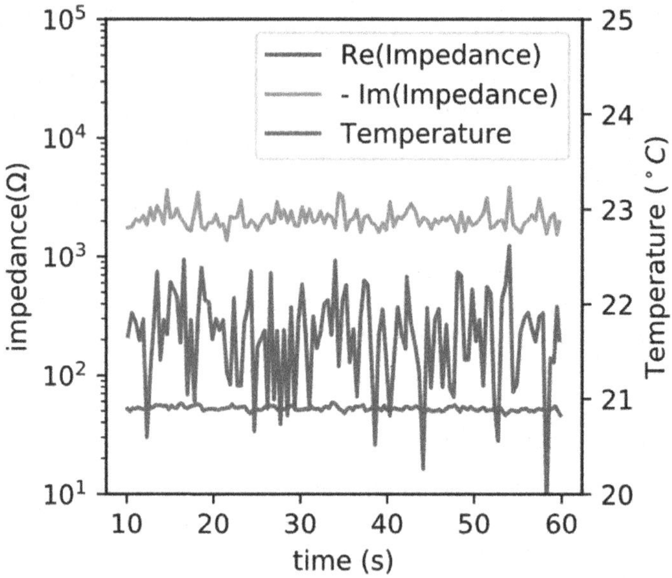

FIGURE 8.4 The measured impedance between contacting surfaces under isothermal condition. Both the real and imaginary part of the measured value did not increase or decrease with time.[19] (Figure use with permission.)

NOTES

1 Kapadia, R., Glyde, R., and Wu, Y., 2007, "In Situ Observation of Phosphorous and Non-Phosphorous Antiwear Films Using a Mini Traction Machine with Spacer Layer Image Mapping," *Tribol. Int.*, 40(10–12), pp. 1667–1679.

2 Xu, Y., Wang, Q., Hu, X., Li, C., and Zhu, X., 2010, "Characterization of the Lubricity of Bio-Oil/Diesel Fuel Blends by High Frequency Reciprocating Test Rig," *Energy*, 35(1), pp. 283–287.

3 Tonck, A., Martin, J. M., Kapsa, P., and Georges, J. M., 1979, "Boundary Lubrication with Anti-Wear Additives: Study of Interface Film Formation by Electrical Contact Resistance," *Tribol. Int.*, 12(5), pp. 209–213.

4 Clarke, A., Weeks, I. J. J., Evans, H. P., and Snidle, R. W., 2016, "An Investigation into Mixed Lubrication Conditions Using Electrical Contact Resistance Techniques," *Tribol. Int.*, 93, pp. 709–716.

5 Jablonka, K., Glovnea, R., and Bongaerts, J., 2012, "Evaluation of EHD Films by Electrical Capacitance," *J. Phys. Appl. Phys.*, 45(38), p. 385301.

6 Schnabel, S., Marklund, P., Minami, I., and Larsson, R., 2016, "Monitoring of Running-in of an EHL Contact Using Contact Impedance," *Tribol. Lett.*, 63(3), p. 35.

7 Jablonka, K., Glovnea, R., and Bongaerts, J., 2012, "Evaluation of EHD Films by Electrical Capacitance," *J. Phys. Appl. Phys.*, 45(38), p. 385301.

8 Jablonka, K., Glovnea, R., Bongaerts, J., and Morales-Espejel, G., 2013, "The Effect of the Polarity of the Lubricant upon Capacitance Measurements of EHD Contacts," *Tribol. Int.*, 61, pp. 95–101.

9 Hadden, T., Jiang, J. W., Bilgin, B., Yinye Yang, Sathyan, A., Dadkhah, H., and Emadi, A., 2016, "A Review of Shaft Voltages and Bearing Currents in EV and HEV Motors," In: *IECON 2016 – 42nd Annual Conference of the IEEE Industrial Electronics Society*, Florence, Italy, pp. 1578–1583.

10 Håkansson, B., Andersson, P., and Bäckström, G., 1988, "Improved Hot-Wire Procedure for Thermophysical Measurements under Pressure," *Rev. Sci. Instrum.*, 59(10), pp. 2269–2275.

11 Shibata, H., Suzuki, A., and Ohta, H., 2005, "Measurement of Thermal Transport Properties for Molten Silicate Glasses at High Temperatures by Means of a Novel Laser Flash Technique," *Mater. Trans.*, 46(8), pp. 1877–1881.

12 Carey, A. A., 1938, *The Dielectric Constant of Lubrication Oils*. Research Report No. 9, The Pennsylvania State College Engineering Experiment Station, State College, PA.

13 Chen, Y., P. A. Renner, and H. Liang, 2021, "Using Electrochemical Impendence to Characterize Thermal Performance of Working Lubricants," *Surf. Topogr. Metrol. Proper.*, 9(3), p. 035035.

14 Turnbull, R., et al., 2023, "Electrotribodynamics of Ball Bearings in Electrical Machines," *Tribol. Int.*, 188, p. 108817.

15 Arole, K., et al., 2023, "Effects of $Ti_3C_2T_z$ MXene Nanoparticle Additive on Fluidic Properties and Tribological Performance," *J. Mole. Liq.*, 386, p. 122435.

16 Guo, L., et al., 2023, "Study on the Electric Discharge Behaviour of a Single Contact in EV Motor Bearings," *Tribol. Int.*, 187, p. 108743.

17 Stribeck, R., 1902, "The Key Qualities of Sliding and Roller Bearings," *Z. Ver. Dtsch. Ing.*, 46(38), pp. 1342–1348.

18 Chen, Y., and Liang, H., 2020, "Tribological Evaluation of Electrical Resistance of Lubricated Contacts," *ASME J. Trib.*, 142(11), p. 114502. doi:10.1115/1.4045578.

19 Chen, Y., Renner, P., and Liang, H., 2021, "Using Electrochemical Impendence to Characterize Thermal Performance of Working Lubricants," *Surf. Topogr: Metrol. Prop.*, 9(3), p. 035035.

20 Tajedini, M., Bahrami, R., Azhdari, R., Sue, H. J., and Liang, H., 2024, "Influence of Electric Current on Tribological Performance of Grease-Lubricated Steels," *Trib. Int.*, 200(12), 110121. doi:10.1016/j.triboint.2024.110121.

9 Adhesion and Tackiness of Greases

Emmanuel P. Georgiou, Dirk Drees, Lais M. Lopes, Michael De Bilde, and Michael Anderson

9.1 INTRODUCTION

Greases are extensively used in various and diverse industrial sectors, including food, transportation, aeronautics, construction, mining, and metallurgy [1]. Their fundamental purpose is to reduce frictional forces and protect industrial components from wear and corrosion damage [2]. Their functionality and performance in the field are inseparably connected to both their interaction properties, such as adherence to a surface, and to some intrinsic characteristics such as cohesion or consistency, and tackiness [3].

Tackiness is considered as the ability of a grease to form threads when being pulled apart and thus plays a vital role in determining the transferring of a grease between contacting areas [4], Figure 9.1. Depending on the application, different specifications and requirements on tackiness are needed. For example, in industries like machinery, it's crucial to have grease with tackiness. This helps the grease transfer easily between surfaces, in order to efficiently lubricate them and keep things running smoothly. Conversely, in a sector like food processing, it's important for

FIGURE 9.1 Transferring of grease between gears in an FZG configuration. Notice the formation of grease threads between the moving gear teeth.

 DOI: 10.1201/9781003496069-9

a grease to have low tackiness. This prevents it from transferring too much and potentially contaminating the products that are being processed. Apart from grease transferring between moving components, it is also essential that a sufficient amount of grease adheres to the surface to maintain lubrication and protect from wear phenomena. Thus, both adhesion and tackiness play an important role in determining the behaviour and performance of the grease.

To complicate things even more, temperature and speed are widely known to affect the functionality of greases in a mechanical system, due to their strong impact on the physical and chemical characteristics of the lubricant [5]. Elevated temperatures can lead to thermal degradation of the grease, altering its viscosity and structural integrity, thereby compromising its lubricating ability. On the contrary, low temperatures may cause the grease to solidify, limiting its flow and hindering proper lubrication. In addition, the speed at which machinery operates affects the distribution and retention of grease within the lubrication contacts. This means that depending on the 'range' of working conditions, different requirements in terms of adhesion and tackiness may be needed. This amplifies the difficulty in the development of new grease formulations and the complexity in the selection of commercially available products by the end-users.

Despite the widespread use of greases and the widely recognized link between adhesion and tackiness with grease functionality, quantification and understanding of these systemic (depend on contacting surface attributes) and intrinsic (depend on formulation) grease characteristics remains a challenge. Therefore, this chapter provides a clear background on the concept of adhesion and tackiness, outlook on existing measurement techniques, influencing contact parameters on adhesion and application examples. As a starting point, the main terms are briefly clarified to explain the basic differences and the context in which the measurement techniques were developed for.

9.2 ESTABLISHMENT OF A TERMINOLOGY

Despite being identified as early as the 1940s [6], the concepts of adhesion and tackiness in greases still lack comprehensive understanding. Indeed, there is a common misconception that the pull-off force, which relates to the ability of a grease to adhere to a surface, is also a direct indication of its tackiness [7]. This typically leads to a confusion between 'tacky' and 'sticky' greases. High pull-off force is a property of a sticky grease. Therefore, a basic terminology needs to be established:

- **Adhesion**: It is identified as the force needed to initiate separation between the two surfaces that are connected by a grease layer. It is considered as a systemic property as it strongly depends on the characteristics of the surfaces (composition, topography, geometry of contact, etc.).
- **Pull-Off Force**: It is the maximum force that must be overcome to separate the two grease-covered surfaces. This is typically considered as the quantification/reference value of the adhesion of a grease in the majority of laboratory setups.

- **Tackiness**: It is considered as the ability of a grease to form (long) threads when being pulled apart between two surfaces. This ability strongly depends not only on the grease itself (intrinsic characteristic) but also on the speed of separation, the temperature and possible other factors.
- **Thread Length**: It is the maximum length of the thread that is being formed during the separation process.
- **Cohesion**: It is identified as the intermolecular attractive forces within a grease. It is considered as a crucial intrinsic characteristic, as it determines the consistency and ability of a grease to maintain its structure under various operating conditions.
- **Separation Energy**: It is defined as the energy needed to fully separate two surfaces that are connected with a grease layer, and it can be a combination of the properties of tackiness and pull-off force.

9.3 CURRENT STATE ON TACKINESS AND ADHESION EVALUATION

The growing complexity of greases [8] presents a significant challenge for both developers and end-users, as it is increasingly difficult to distinguish between available options in research or the market. Consequently, selecting the most suitable grease for a specific application becomes a daunting task, since there is no standardized test procedure available. Up to date, there are several test methods developed for measuring the adhesion and tackiness of greases. Some of them are purely empirical, whereas in recent years an attempt is made to develop more sophisticated methods that can quantify the adhesion force, cohesion and thread formation (tackiness). Hereafter, an overview of the most widely used test procedures is given:

- **Finger Test**: Until now, the finger test is one of the most used approaches to easily assess the tackiness of greases qualitatively [9]. This empirical method involves applying a layer of grease between the thumb and index fingers and then pulling them apart, resulting in the formation of grease threads, Figure 9.2. In accordance with this test, longer threads indicate higher tackiness of the grease; for example in Figure 9.2, the grease on the right side forms longer threads, and thus, it is 'tackier'. While the finger test offers a simple and cost-effective method, it is also highly empirical and subjective, as the results heavily rely on the individual performing the test. Factors such as the amount of grease used and the speed of retraction can lead to varying results. Despite its inherent subjectivity, the grease finger test remains a valuable tool in the array of grease evaluation techniques, providing practical, hands-on insights that complement more complex analytical methods.
- **Hammer Test**: Another field test that is also reported to assess adhesion and tackiness of greases is the 'hammer' test [10]. In this method, a five-pound hammer head is dropped onto a small quantity of grease applied on an anvil. The distance to which the grease is spread determines its tackiness (higher

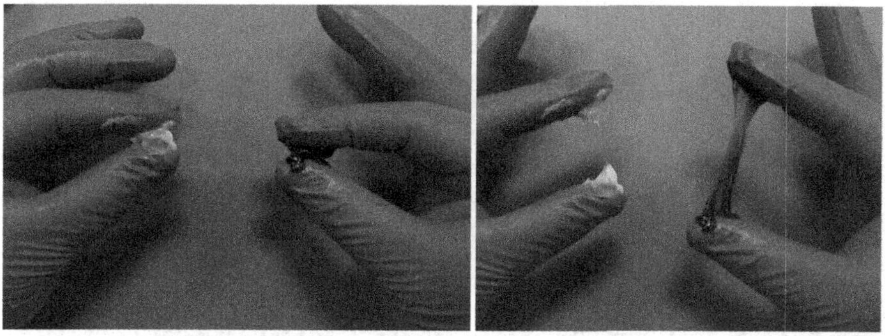

FIGURE 9.2 Comparison of two commercially available greases via the finger test.

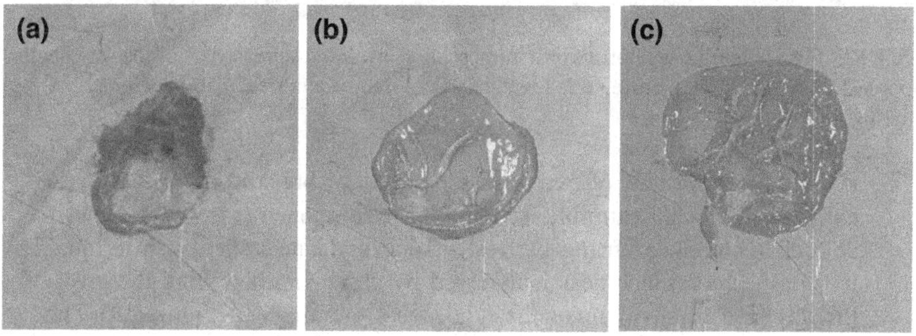

FIGURE 9.3 Comparison of greases via the hammer approach (same scale). Notice that the tackiness (spread of grease) increases in the following order (a) < (b) < (c). Also, there is a different amount of grease remaining in the middle of the initial grease pot, which is an indication of different adhesion.

distance means higher tackiness). In addition, the adhesion can be subjectively assessed by evaluating the amount of grease that remains in the initial area. This method has the disadvantage that it is not well controlled and that it measures more the cohesion of the grease upon impact rather than quantifying the adhesion or tackiness. In addition, the sensitivity of this approach is rather poor, especially when comparing greases with similar consistency (NLGI number). But in any case, the hammer test may serve as a practical means of simulating contact scenarios where greased components may experience sudden shocks or impacts (Figure 9.3).

- **Eggbeater Test**: This is yet another simple and practical method [11] that is used in the field for a practical assessment of the cohesion and tackiness of a grease. Based on this procedure, a clean and dry handheld eggbeater or a similar mixing tool is inserted in a pot of grease. Then the eggbeater is turned on, mixing the grease for a specific duration. The duration can vary depending on the end user, but it typically lasts for just a few minutes. Subsequently, the cohesion of the grease is evaluated by observing

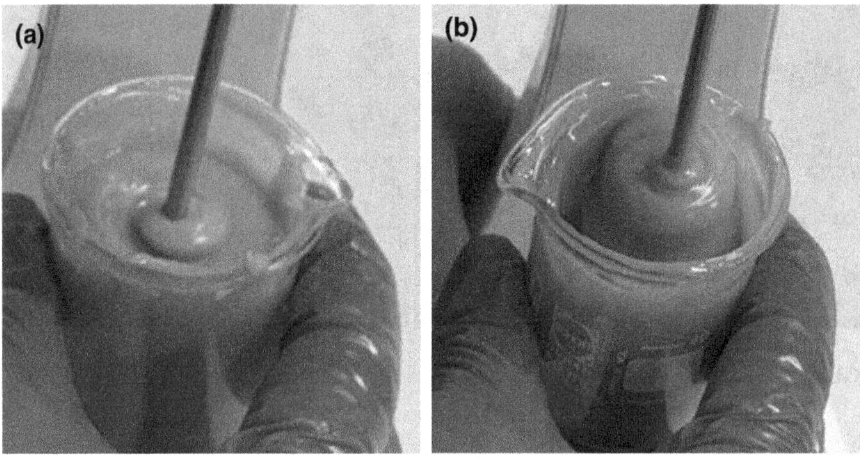

FIGURE 9.4 Comparison of two commercially available greases via the eggbeater approach. The grease in picture (a) is less tacky than that of picture (b), it does not climb the stirring axis as much.

if there are any signs of excessive separation or breakdown. If the grease exhibits significant thinning, separation of components, or other undesirable characteristics during the test, it may indicate issues with its quality. Furthermore, tackiness is assessed by observing how high the grease threads have 'climbed' up on the eggbeater's spinning rod (Figure 9.4). This phenomenon is based on the 'Weissenberg effect' that occurs when a spinning rod is inserted into an elastic or non-Newtonian liquid, where this liquid is drawn towards the rod and rises up around it instead of being thrown outward. Again, it's important to note that this test is a qualitative method and does not provide any form of quantitative data. It can serve as a quick and effective means of assessing grease consistency and thread formation under mixing conditions.

- **Pull-Off Test**: The first test developed that could quantify the adhesion of greases was the pull-off test [12]. Its working principle involved pressing two surfaces together with a controlled amount of grease, until reaching a predetermined load (Figure 9.5a). By doing so, a relatively uniform grease layer with no air pockets was formed between the two surfaces. The load was maintained for a few seconds, and then, the two surfaces were pulled apart (Figure 9.5b); the maximum load needed for complete separation of the two surfaces was recorded (Figure 9.5c). This load was considered as the pull-off force, which related to the adhesive strength of the grease layer. In recent years, updated laboratory versions and commercial products based on this method emerged [13], which also allow to measure the length of formed threads. This is achieved either by adding a camera or an extensometer. In addition, heating of the plates was also added to investigate the effect of temperature on the adhesion and tackiness of a grease. However, existing pull-off setups [13] utilize load cells in the range of 50–100 N

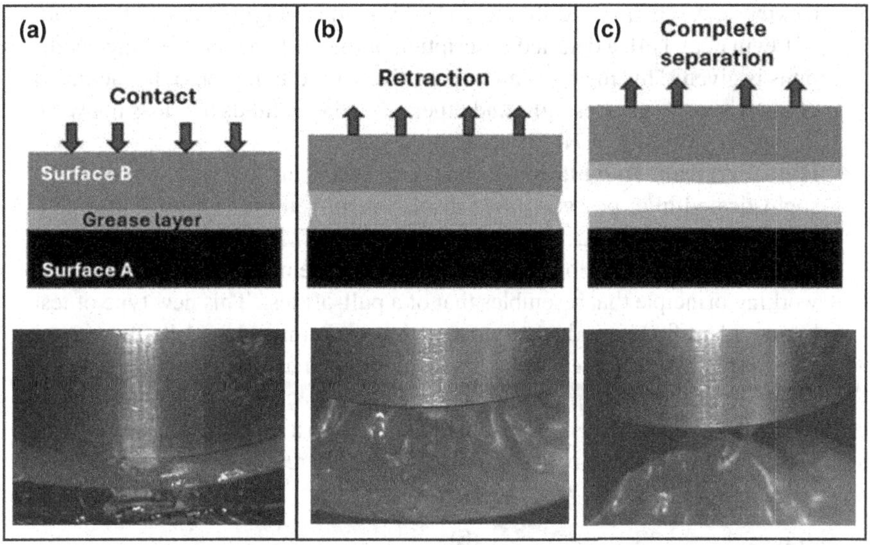

FIGURE 9.5 Three main steps of a pull-off experiment, from left to right. Surfaces with a grease layer come into contact; surfaces are pulled apart while the grease layer deforms forming threads; and the test ends when complete separation occurs.

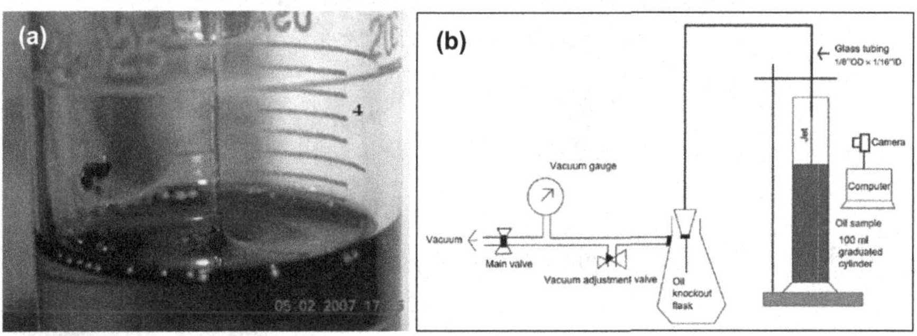

FIGURE 9.6 (a) Formation of a thread during an open siphon test [14]. (b) Schematic of open siphon setup [14].

(precision around 0.1 N), so it could be difficult to differentiate between greases with similar formulation.

- **Open Siphon Method**: is a technique used to assess the flow properties of greases and liquids under the influence of vacuum. In particular, a specific volume of grease is placed into a container with an open-ended capillary tube inserted vertically into the grease. Vacuum acts on the grease, causing it to flow through the capillary, as shown in Figure 9.6a. The rate of flow over a certain period is measured with a camera, providing information about the viscosity and flow characteristics of the grease. In addition, the maximum length of the thread that is formed is recorded to give an indication on tackiness. A schematic of this setup [14] is given in Figure 9.6b. In this work of Levin et al. [14], a detailed description of this test and the working conditions is given. This method can be effectively used to measure the tackiness of low viscosity greases, oils, and other viscoelastic fluids but does not work for higher viscosity or NLGI grade products.

- **High Precision Indentation-Retraction Test**: During the last few years, sophisticated high precision load-displacement sensors (100 mN load cell, 1/10.000 resolution) have become available [3,4,7,15]. These sensors were incorporated in adhesion and tackiness measurements (Figure 9.7a), in a working principle that resembles that of a pull-off test. This new type of test is now identified as indentation-retraction experiments, and its theoretical principle is based on measuring the interaction forces between a probe and a grease layer with high precision on force and retraction distance. According to this method, a fixed grease volume is applied on a standardized holder with multiple small reservoirs that are filled with the grease sample, Figure 9.7b.

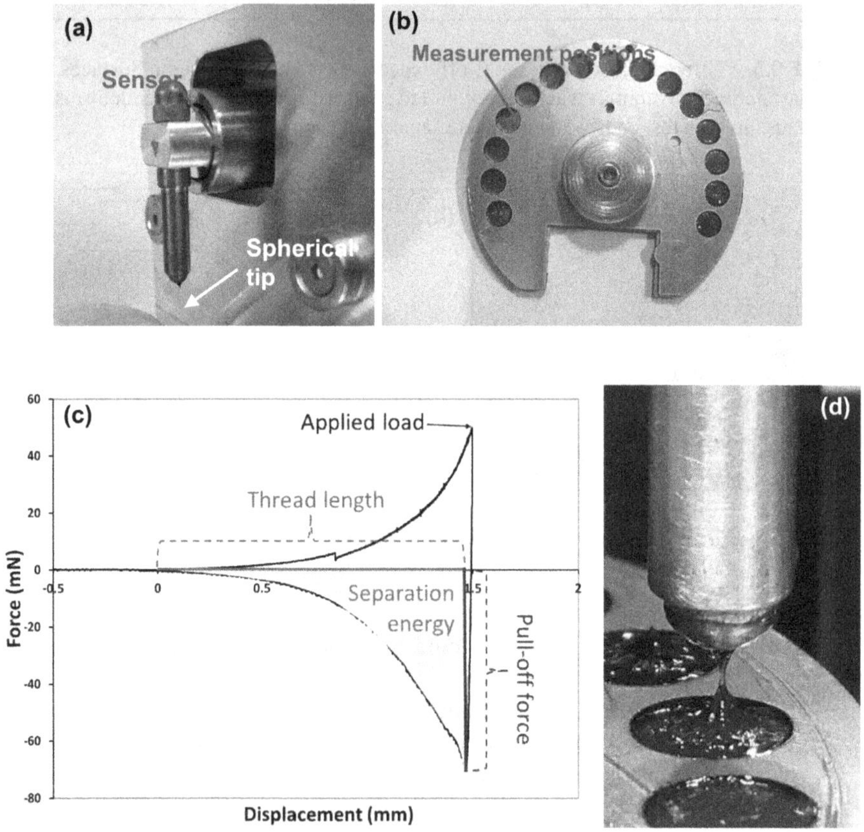

FIGURE 9.7 (a) Indentation-retraction sensor and probe. (b) Grease holder for indentation-retraction experiments. (c) Typical indentation-retraction curve. (d) Formation of a single thread during an indentation retraction test.

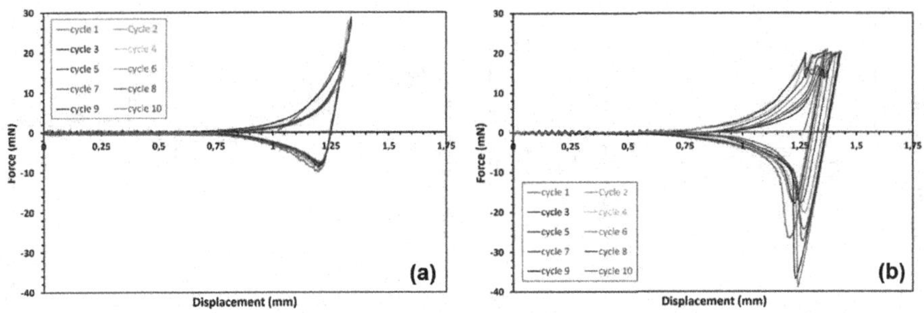

FIGURE 9.8 Example of ten indentation-retraction measurements for a grease with (a) a spherical tip and (b) with a flat tip. Notice the fluctuation of curves for the latter case.

An indenter having a spherical counter-material is gradually brought into contact with the grease layer and pressed down with a preset load (applied load in Figure 9.7c). The load is maintained for a dwell time (typically between 5 and 10 s) until a homogeneous film/layer with no air pockets is formed at the contact area between sphere and grease holder. Then, the indenter is gradually moved away from the greased substrate under well-controlled conditions until there is complete physical separation (Figure 9.7c). By selecting a spherical probe, during the retraction process only a single thread is formed (Figure 9.7d) in a repeatable pattern, significantly improving repeatability of the test method. Indeed, as can be seen from Figure 9.8, a significantly higher fluctuation in the morphology of indentation-retraction curves (repeats) is observed when using a flat (area contact) instead of a spherical tip on the probe (Figure 9.6).

Based on this indentation-retraction approach, the pull-off force is identified as the force required to start the physical separation of the indenter from the grease (indicated by minimum force in Figure 9.7c). As the indenter is retracted, a grease thread forms, which exerts a non-zero force on the force sensor. The thread length is defined as the displacement from the start of the separation (pull-off force) until the sensor force is zero (thread is broken). In addition, the separation energy (Se) is defined as the energy needed to fully separate the indenter from the grease and is calculated by integration of the area between the start of the retraction (pull-off force) and complete physical separation (Figure 9.7c). Based on this method, tackier greases form longer threads (property indicated by the finger test), whereas stickier greases display a higher pull-off force (relates to the adhesion). Separation energy can be used as an additional criterion for comparison of greases, as it combines adhesive, cohesive and tackiness characteristics. This test method allows to change the retraction speed and heat the grease to see variations in its behaviour. Despite the fact that in the holder (Figure 9.7a) there are multiple test locations (reservoirs), it is strongly advised that one grease is tested at a time to avoid cross contamination. Additionally, it is also recommended to 'work' the grease with a micro-grease worker (recommended: 60 strokes for 1 min – 1 stroke per sec), to homogenize it before testing. This protocol greatly improves repeatability and reproducibility.

An indicative example of the repeatability of this method in terms of pull-off force and thread length between different users and instruments is presented in Figure 9.9. It was reported [7] that the fluctuation within one test group can be in

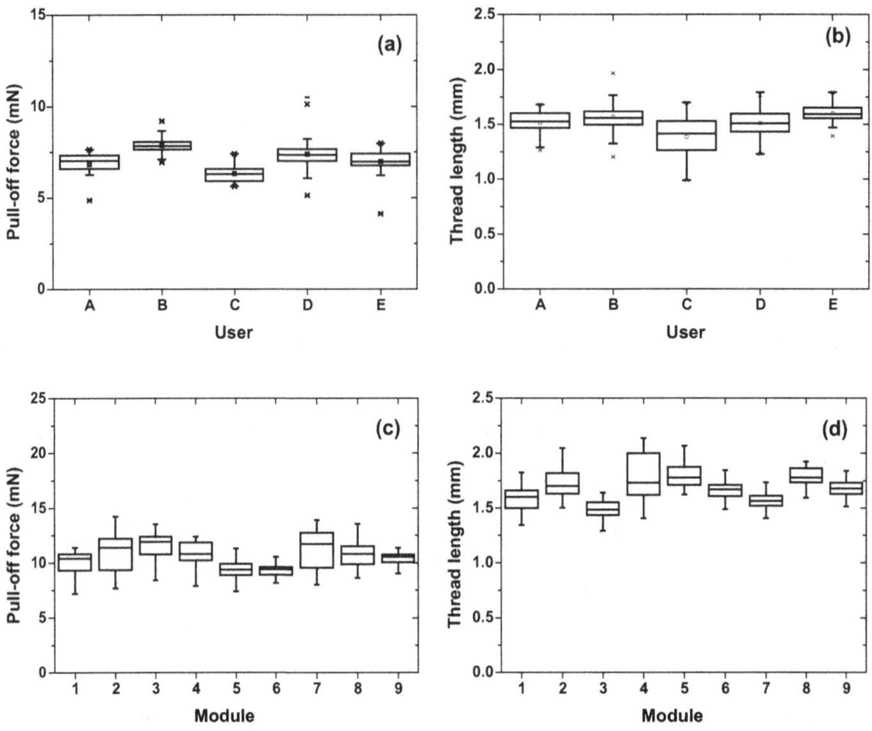

FIGURE 9.9 Repeatability of pull-off force for retraction speed of (a) 0.1 and (c) 1 mm/s at 30°C. Repeatability of thread length for retraction speed of (b) 0.1 and (d) 1 mm/s at 30°C.

the range of 8%–16% for the pull-off force and 4%–12% for the thread length and between the different instruments around 8% and 7%, respectively, at the lower retraction speeds (0.1 mm/s). When the retraction speed increased (1 mm/s), the fluctuation within one test group was in the range of 4%–25% for the pull-off force and 6%–12% for the thread length and between the different instruments 16% and 11%, respectively. The observation that the fluctuation between tests (one test group, one instrument) and between instruments, in other words, the ASTM Repeatability and Reproducibility are hardly different, indicates that this test method does not depend on the instrument and little on the operator, fluctuations measured are mostly to be attributed to the inhomogeneity of a grease property. Since August 2024, the indentation-retraction method has become a standardized method (ASTM D 8576-24) for measuring the thread length and pull-off force of lubricating greases.

9.4 FACTORS INFLUENCING THE ADHESION AND TACKINESS OF GREASES

It is widely known that the performance of greases is strongly dependent on the conditions under which they operate. Indeed, environmental factors such as temperature and humidity, as well as the contact conditions including motion, contact pressure, speed, etc., affect their functionality in the field. Therefore, to select the appropriate grease formulation that meets the specifications of the end user application, the test conditions should match as closely as possible those in the field. The adhesive behaviour and tackiness of greases are also influenced by these factors. Hereafter, indicative examples measured with the indentation-retraction method (aka TAA method, for Tackiness and Adhesion Analyser) are given:

9.4.1 Effect of Tackifier Content on Adhesion and Tackiness

A tackifier is an additive commonly used in lubricants to provide a sticky and/or stringy (tacky) quality. It's primary function is to enhance adherence of the lubricant to surfaces and promote thread formation for grease transferring between moving components. Tackifiers serve to prevent dripping, displacement, or splattering of oils, and they can also contribute to the texture of greases. However, to optimize the performance of a grease, the correct selection of tackifiers and concentration is crucial. For example, in Figure 9.10, a comparison of different commercially available tackifiers A-H (1% concentration) shows a different influence on the base grease under the same test conditions (2 mm/s retraction speed, 25°C, 3 mm ø copper spherical probe). Some tackifiers improved the ability of the base grease to form threads (A, C and H), whereas others resulted in no improvement or even a reduction of the tackiness (e.g. tackifier G). In addition, it appears that tackifiers do not influence the adhesion in the same way: tackifier H improved the tackiness but did not improve the adhesion of the base grease. Thus, it appears that adhesion and tackiness are two different grease properties/characteristics that should not be directly correlated.

Additionally, when the correct combination of base grease and tackifier is achieved, the increase of tackifier content typically results in the increase of thread formation (tackiness), Figure 9.11. However, a reverse correlation with the adhesion is present here. This indicates that by increasing tackifier content, the grease can transfer easily between moving components but there is a growing risk that it can drip away from the initial contact. Thus, adhesion and tackiness should be determined upon the requirements of the application.

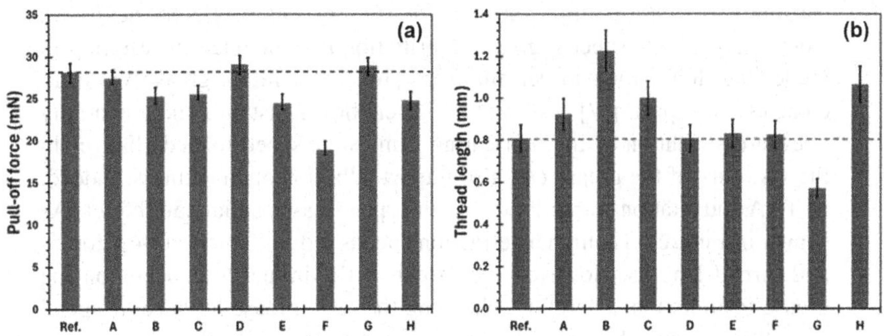

FIGURE 9.10 Effect of different tackifiers on the (a) pull-off force and (b) thread length of a grease. Notice the dissimilar influence (tackifier content: 1%, test conditions: 2 mm/s retraction speed, 25°C, 3 mm ø copper spherical probe).

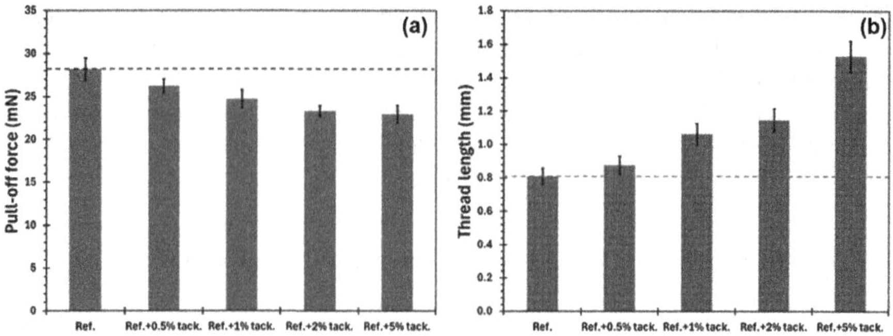

FIGURE 9.11 Effect of tackifier content on the (a) pull-off force and (b) thread length of a grease (test conditions: 2 mm/s retraction speed, 25°C, 3 mm ø copper spherical probe).

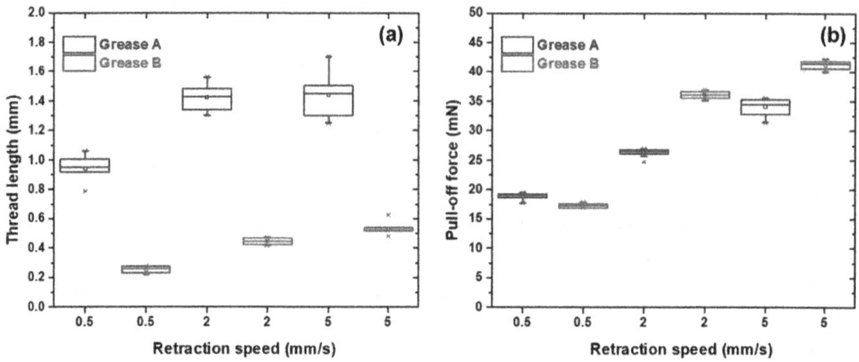

FIGURE 9.12 Effect of retraction speed on the (a) pull-off force and (b) thread length of two commercially available greases (test conditions: 25°C, 3 mm ø copper spherical probe).

- **Effect of Speed**: Speed significantly influences grease performance in mechanical applications. High speeds can generate heat through friction, causing the grease to thin out and lose its lubricating effectiveness [16]. Conversely, at low speeds, grease distribution may be uneven, resulting in inadequate lubrication in certain areas [16]. In addition, grease viscosity changes with speed [17], leading to different flow resistance and potentially uneven distribution of the lubricating film. Such speed-related changes in the viscosity of the grease can also impact adhesion and tackiness. Indeed, in TAA indentation-retraction tests, the speed has a significant effect. As shown in Figure 9.12, higher retraction speeds led to higher pull-off forces and thread lengths. However, the extent of the influence depends on the grease formulation. Therefore, choosing the right grease for specific speed conditions is crucial to ensure optimal performance and prolong component lifespan.
- **Effect of Temperature**: Temperature is a key parameter in grease functionality as it influences the viscosity and consistency and has a strong effect on

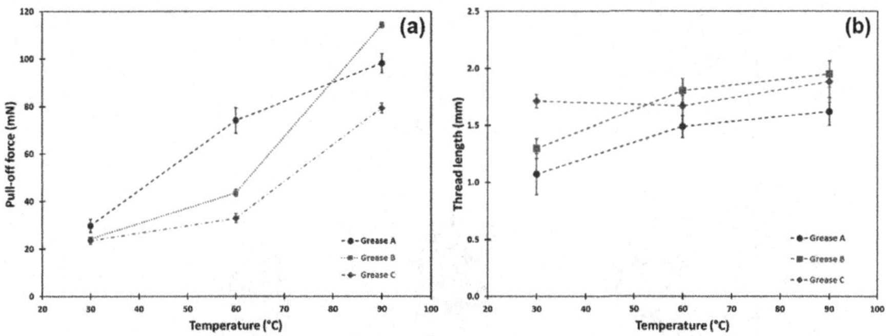

FIGURE 9.13 Effect of temperature on the (a) pull-off force and (b) thread length of commercially available greases (test conditions: 1 mm/s retraction speed, 3 mm ⌀ copper spherical probe).

grease performance and lifespan [18]. Recent work [4,7,15] has shown that it also influences adhesion and tackiness. As shown in Figure 9.13, different trends emerge, depending on the grease formulation and temperature. For instance, grease B and C exhibit a notable increase in pull-off force at 90 oC, indicating higher adhesion. This increase appears to be more gradual for grease A. When assessing the tackiness of these same greases, the increase of temperature resulted in longer thread formation for greases A and B. On the contrary, grease C appears less temperature-dependent (within the chosen range). Also in this case, it's crucial to notice that the effect of temperature differs for adhesion or tackiness. Similar temperature and pull-off force correlations were observed for other commercial greases and speeds (0.1 up to 5 mm/s), which clearly indicated that adhesion and tackiness are both temperature-related but not necessarily correlated. Indeed, temperature variations significantly influence the viscous and viscoelastic responses of greases [19]. Moreover, phase transitions within the grease's microstructure can impact its yield strength [20] and the force necessary for its deformation and separation.

- **Effect of Applied Load**: For the indentation-retraction method, changes in applied load had a small effect on adhesion and tackiness, Figure 9.14. This was linked to the ball-on-flat configuration that results in the formation of single threads and to reduced load fluctuations (e.g. brought about by vibrations) [15].
- **Effect of Film Thickness:** In the field, it is known that the quantity of grease used can influence the lubricity of the tribo-contact and the overall performance of the system. Thus, it is important to identify whether adhesion and tackiness are also influenced. Indentation-retraction measurements have shown (Figure 9.14) that the initial grease layer thickness has little effect on the measurement, at least in the range of applying 100–500 μm [15]. This is because the excess of grease is 'pushed away' from the tribo-contact during the indentation phase, and what truly determines the actual thickness of the grease in these experiments is the characteristics of the countermaterial and the contact conditions.

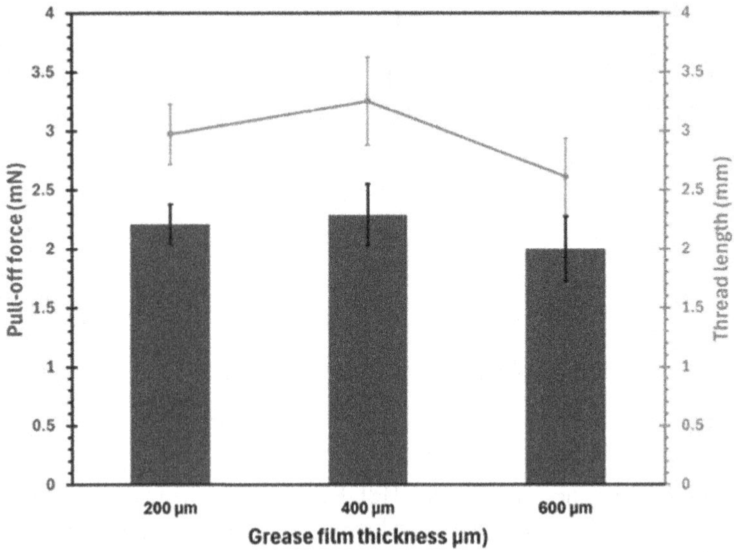

FIGURE 9.14 Effect of film thickness on the (a) pull-off force and (b) thread length of a commercially available grease (test conditions: 1 mm/s retraction speed, 3 mm ø copper spherical probe).

- **Effect of Surface Characteristics**: Since adhesion is a surface phenomenon, the surface roughness/topography of the contact materials should be considered. In terms of roughness, it was found that surface roughness in general does not have a significant effect on adhesion and tackiness. The only noticeable difference is a higher spread of measurement results at the highest roughness of the surface.

On the contrary, material selection can have an influence on adhesion. For example, by changing the probe composition [15] and/or adding an anti-stick surface [7], the adhesion is strongly influenced. However, maybe surprisingly, the tackiness was not significantly influenced by the composition of the contacting surfaces, Figure 9.15. The ability of a grease to form threads seems to be more an intrinsic characteristic, rather than related to the grease adhesion that appears to be a systemic property.

Finally, as mentioned before, contact geometry has a strong effect on pull-off force and tackiness (Figure 9.8). The larger the contact, the more threads that are being formed, and this leads to more variability.

9.5 CORRELATING ADHESION AND TACKINESS TO TRIBOLOGICAL PERFORMANCE

With the advancement of measuring techniques, a link between adhesion and tackiness of greases with their frictional performance is currently being explored. The general perception is that friction in a tribo-contact is determined by adhesion and cohesion that are the two mechanisms driving grease flow at the contacting

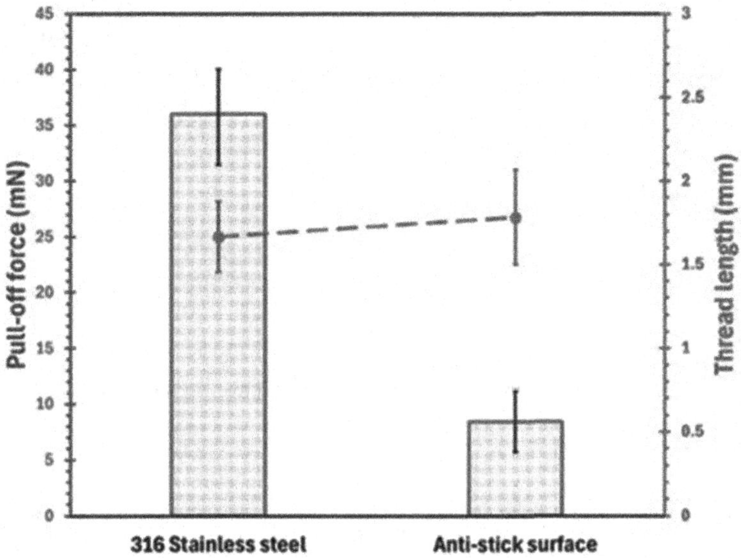

FIGURE 9.15 Effect of surface on the (a) pull-off force and (b) thread length of a commercially available grease (test conditions: 1 mm/s retraction speed, 3 mm ⌀ copper spherical probe).

interface [21]. Adhesion controls how well the grease will remain in the initial surface, cohesion regulates the shearing of the grease that leads to its ejection from the tribo-contact and tackiness dictates the transferring of grease between surfaces due to thread formation. These interactions are strongly dependent on the motion under which the surfaces are interacting [22]. Indeed, in sliding motions, thread formation plays a minor role, whereas the behaviour of the grease is affected mainly by its adhesion and cohesion that determine the amount of grease that stays in the tribo-contact, as shown in the work of Ferrer et al. [23]. In accordance with this work, Georgiou et al. [24] observed that when testing greases with different concentrations and types of tackifiers, Figure 9.16 under reciprocating sliding conditions, the tackiness doesn't significantly affect friction. Even when testing at different sliding frequencies, the only noticeable differences were observed at the initial stage of the test and at lower frequencies (speeds). This variance at lower frequencies can also be due to the homogenization of the grease layer at the tribo-contact. In addition, for the same type of reciprocating sliding tests, a similar behaviour was also observed by changing the type of tackifiers, which is an indication that tackiness does not play a vital role in these reciprocating sliding conditions.

On the contrary, tackiness seems to play an important role under rolling/sliding conditions, as higher tackifier content resulted in a lower friction in a traction experiment, Figure 9.17a. However, based on existing know-how a direct relationship between higher tackiness and lower friction cannot be made at the moment. In addition, the combination of different tackifiers with the same base oils can have a different impact on the friction performance of the system (Figure 9.17b). Thus,

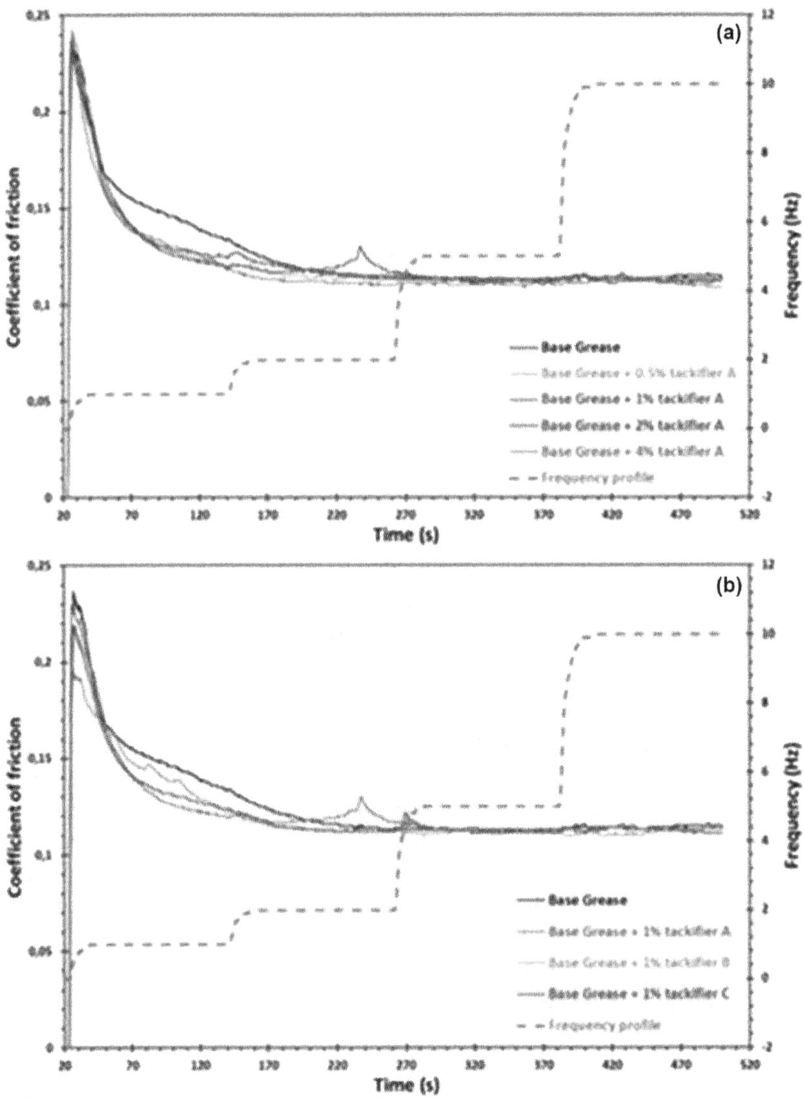

FIGURE 9.16 Effect of (a) tackifier concentration and (b) different types of tackifiers on the sliding friction of a base grease tested under reciprocating sliding conditions (DIN 6325 steel plate, 6 mm Ø x 6 mm length Cr6 cylinder, 150 N applied load, 500 MPa Hertzian pressure).

other characteristics in grease formulation should also be considered. In any case, it is believed that during testing of greases under rolling conditions both adhesion and tackiness are important factors. Adhesion determines the amount of grease left on the disk (Figure 9.18a), while tackiness governs grease transfer from disk to ball via thread formation (Figure 9.18b).

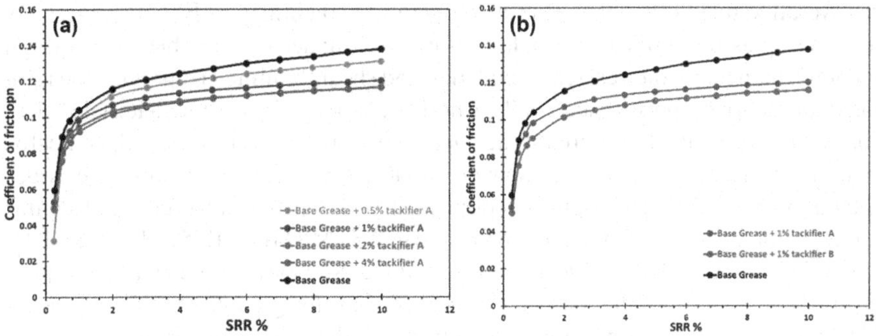

FIGURE 9.17 Effect of (a) tackifier concentration and (b) different types of tackifiers on the friction of a base grease tested under rolling/sliding conditions in a traction experiment (AISI 52100 disks and balls, 20 N applied load, 1000 mm/s speed).

FIGURE 9.18 Formation of grease threads on the (a) ball and (b) disk after an MTM test.

9.6 EXAMPLES OF ADHESION AND TACKINESS MEASUREMENTS IN OTHER INDUSTRIAL SECTORS

In industrial applications, adhesion and tackiness of greases and lubricants play pivotal roles in ensuring efficient lubrication and prolonged equipment longevity. Adhesion is essential for the ability of the lubricant to attach onto the component's surface and to form a film, cohesion determines its resistance to deformation under pressure or shear forces, and tackiness relates to its capacity to transfer between moving surfaces via thread formation. Thus, the selection of greases and lubricants with optimal adhesion, cohesion and tackiness characteristics is crucial for maintaining equipment reliability, minimizing maintenance downtime, and extending component service life in diverse industrial environments.

In the *automotive industry*, grease development and optimization are gaining extreme importance as electric vehicles (EVs) are emerging as the future of transportation and the market shifts away from internal combustion engine vehicles [25]. In Evs, grease is used in critical components such as rolling bearings [26] to reduce friction and wear, and thus ensuring a smoother component operation and enhancing the overall vehicle efficiency. However, operating conditions in EVs are extremely challenging as the lubricants should perform well under the variable speeds (high and low), high temperatures, resist oxidation and shear thinning and provide the right bleed and lubricity performance [27]. Another challenge is that the materials selection for the bearings is different than that used in traditional motors. Specifically, many bearings have ceramic components that act as insulators to mitigate issues related to stray current. Thus, the tribo-chemistry of the contact has changed significantly, while majority of standardized test (e.g. ASTM D4170, D7594 Fretting, DIN 51819 FAG FE8, ASTM D2596-15 and ASTM D2783-15) still use metallic samples. In addition, grease formulation is also modified since the greases should not be electrically conductive, to avoid stray current transfers that lead to bearing damage [28]. This means that current specifications need to change, whereas emphasis should be given on modifying existing test procedures to match as close as possible the conditions of the application. Among the different methods used nowadays for evaluating greases for EVs, adhesion and tackiness is not included even though these characteristics are linked to friction performance, especially in rolling contacts (as shown in paragraph 5). In the review article by Chen et al. [25] the influence of grease formulation on their lubricating properties for EVs is stressed out. From the existing adhesion-tackiness measurements methods the highest variability of test conditions is available with the high precision indentation-retraction method, since changes can be made in the retraction speed, applied temperature and probe/plate composition. Therefore, it can be finetuned to simulate closer the operating conditions and tribo-contact met in EVs.

Recent reports showed [23,29,30] that in the *railway industry* proper grease formulation can optimize lubricity, improve wear resistance, and minimize grease loss due to centrifugal forces. Ferrer et al. [23] investigated the tribological characteristics of fully formulated greases having different combinations of base oils and thickeners. They observed that Lithium greases showed lower friction, but higher wear compared to calcium greases, while higher viscosity reduced wear but increased friction universally. Lithium greases with high base oil viscosity were found to be optimal for wheel-rail interface lubrication. This improved behaviour was linked to the grease pick-up (adhesion) and carry-down (cohesion and tackiness) ability. They also observed that apart from the formulation of the grease, the amount and way of applying the grease onto the rail has an influence.

Lubrication is crucial for the *wind turbine industry*, ensuring smooth operation and longevity by reducing friction and wear in components like bearings and gearboxes [31]. Specially formulated lubricants withstand the harsh conditions turbines face, helping to maintain efficiency and prevent downtime. Regular maintenance, including monitoring and replacing lubricants, is essential for optimal performance and sustainability in renewable energy systems. However, during maintenance and repair procedures, as well as during the restart of wind turbines after prolonged periods of inactivity there's a high risk of components operating under lubricant starvation conditions. Maintaining sufficient lubrication is essential to prevent wear damage and ensure smooth operation during the critical start-stop phases [32]. The ability of a lubricant to 'stay' in the tribo-contact is strongly related to its adhesion (Figure 9.19a) primarily, and its cohesion and tackiness will subsequently determine how well it

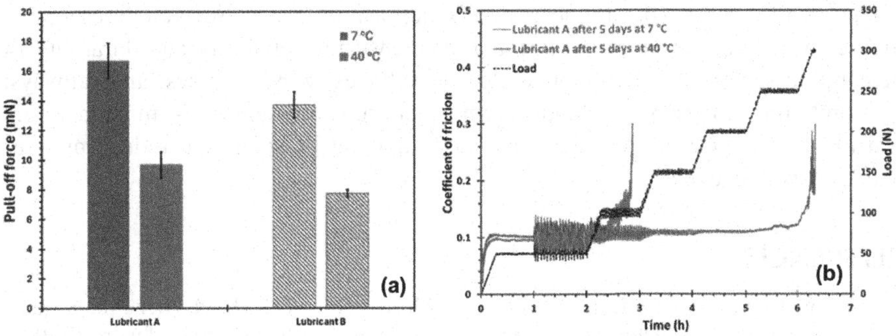

FIGURE 9.19 (a) Effect of temperature on adhesion (pull-off force) of two wind turbine lubricants. (b) Example of evolution of friction after the samples remained inactive in a vertical position for 5 days at a different temperature. Notice the rapid increase of coefficient of friction due to starvation.

will transfer and cover the tribological contact once the motion starts (Figure 9.19b). The use of high precision indentation-retraction methods can be used to determine the adhesive properties of the grease and under variable environmental conditions and test if there is sufficient lubricant remaining in the tribo-contact after a prolonged inactivity duration [33]. This approach can be expanded to any mechanical system that could potentially operate under starvation conditions.

Apart from the transportation industry, adhesion and tackiness are also becoming an increasing topic of interest in the packaging [34,35] and food [36,37] industry, as well as for healthcare and consumer products [38–40].

9.7 SUMMARY

Tackiness and adhesion are critical properties of industrial greases, directly impacting their performance across diverse applications. Assessing these properties has evolved from subjective methods like the finger test, which evaluates tackiness by forming threads between fingers, to advanced laboratory techniques like the indentation-retraction test (TAA). Traditional methods such as the hammer test and eggbeater test provide practical but imprecise evaluations, focusing on grease spread or thread formation. More quantitative approaches, including pull-off tests and the open siphon method, measure specific properties like force separation and flow behavior but are limited in precision for thicker greases. Among modern techniques, the TAA method stands out for its accuracy and repeatability, employing high-precision sensors to measure pull-off force, thread length, and separation energy under controlled temperature and speed conditions. Recently standardized as ASTM D 8576-24, this method minimizes operator influence, enabling reliable grease characterization for demanding industrial applications.

Advanced measurement methods provide valuable insights into how environmental and operational factors—such as temperature, speed, and pressure—affect grease adhesion and tackiness. For instance, while increased tackifier content enhances thread formation, it may compromise adhesion, emphasizing the importance of balancing these properties to meet specific application needs. Such analyses enable grease formulators to optimize formulations, ensuring reliability and durability in demanding sectors like automotive electric vehicles, wind turbines, and railways. By combining traditional techniques with modern evaluation tools, manufacturers can improve grease performance, extend equipment lifespan, and reduce maintenance costs effectively.

REFERENCES

1. Gow, G. (2010). Lubricating grease. In: Mortier, R., Fox, M., & Orszulik, S. (eds) *Chemistry and Technology of Lubricants. Springer*, Dordrecht. https://doi.org/10.1023/b105569_14.
2. Lugt, P. M. (2016). Modern advancements in lubricating grease technology. *Tribology International*, 97, 467–477. https://doi.org/10.1016/j.triboint.2016.01.045.
3. Achanta, S., Jungk, M., & Drees, D. (2011). Characterisation of cohesion, adhesion, and tackiness of lubricating greases using approach–retraction experiments. *Tribology International*, 44(10), 1127–1133. https://doi.org/10.1016/j.triboint.2011.04.019.
4. Georgiou, E. P., Drees, D., De Bilde, M., & Anderson, M. (2018). Can we put a value on the adhesion and tackiness of greases? *Tribology Letters*, 66(2). https://doi.org/10.1007/s11249-018-1011-4.
5. Lugt, P. M. (2009). A review on grease lubrication in rolling bearings. *Tribology Transactions*, 52(4), 470–480. https://doi.org/10.1080/10402000802687940
6. Bikerman, J. J. (1947). The fundamentals of tackiness and adhesion. *Journal of Colloid Science*, 2(1), 163–175. https://doi.org/10.1016/0095-8522(47)90017-2.
7. Georgiou, E. P., Drees, D., De Bilde, M., Anderson, M., Carlstedt, M., & Mollenhauer, O. (2021). Quantification of tackiness of a grease: The road to a method. *Lubricants*, 9(3), 32. https://doi.org/10.3390/lubricants9030032.
8. Rawat, S. S., & Harsha, A. P. (2019). Current and future trends in grease lubrication. In: Katiyar, J., Bhattacharya, S., Patel, V., & Kumar, V. (eds) *Automotive Tribology. Energy, Environment, and Sustainability.* Springer, Singapore. https://doi.org/10.1007/978-981-15-0434-1_9.
9. Gay, C., & Leibler, L. (1999). Theory of tackiness. *Physical Review Letters*, 82(5), 936–939. https://doi.org/10.1103/physrevlett.82.936.
10. Moon, M. (2018). On the right tack. *Lubes'n'Greases.* https://www.lubesngreases.com/magazine/24_10/on-the-right-tack/ (acceced 25 November 2024).
11. Lugt, P. M. (2012). Grease lubrication in rolling bearings. https://doi.org/10.1002/9781118483961.
12. Vargo, D. (2014). The adhesiveness of grease. In: *Proceedings of the NLGI 81st Annual Meeting*, Palm Beach Gardens, FL, USA, 14–17 June 2014.
13. Harmon, M., Powell, B., Barlebo-Larsen, I., & Lewis, R. (2019). Development of grease tackiness test. *Tribology Transactions*, 62(2), 207–217. https://doi.org/10.1080/10402004.2018.1526991.
14. Levin, V. J., Stepan, R. J., & Leonov, A. L. (2009). Evaluating tackiness of polymer-containing lubricants by open siphon method: Experiments, theory, and observations. In: Rudnick, L. R. (ed) *Lubricant Additives Chemistry and Applications,* 2nd edition. Springer, Singapore, 2009, 357–375 (https://doi.org/10.1201/9781315120621)
15. Georgiou, E. P., Drees, D., De Bilde, M., & Anderson, M. (2019). Quantitative approach to measuring the adhesion and tackiness of industrial greases, *Tribologie und Schmierungstechnik,* 66 (2019) 37–43.
16. Chi, J.-B., Ma, H.-J., Lee, G.-S., Park, J.-H., & Kim, D.-E. (2024). Effect of speed and number of sliding cycles on the film formation behavior of fluorine-based grease. *Lubricants*, 12(1), 26. https://doi.org/10.3390/lubricants12010026.

17. Cen, H., Lugt, P. M., & Morales-Espejel, G. (2014). On the film thickness of grease-lubricated contacts at low speeds. *Tribology Transactions*, 57(4), 668–678. https://doi.org/10.1080/10402004.2014.897781.
18. Shah, R., Tung, S., Chen, R., & Miller, R. (2021). Grease performance requirements and future perspectives for electric and hybrid vehicle applications. *Lubricants*, 9(4), 40. https://doi.org/10.3390/lubricants9040040.
19. Salomonsson, L., Stang, G., & Zhmud, B. (2007). Oil/thickener interactions and rheology of lubricating greases. *Tribology Transactions*, 50(3), 302–309. https://doi.org/10.1080/10402000701413471.
20. Zhou, C., Ren, G., Fan, X., & Lv, Y. (2022). Probing the effect of thickener microstructure on rheological and tribological properties of grease. *Journal of Industrial and Engineering Chemistry*, 111, 51–63. https://doi.org/10.1016/j.jiec.2022.03.010.
21. Westerberg, L. G., Höglund, E., Lugt, P. M., Li, J., & Baart, P. (2015). Free-surface grease flow: Influence of surface roughness and temperature. *Tribology Letters*, 59(1). https://doi.org/10.1007/s11249-015-0537-y
22. De Laurentis, N., Kadiric, A., Lugt, P., & Cann, P. (2016). The influence of bearing grease composition on friction in rolling/sliding concentrated contacts. *Tribology International*, 94, 624–632. https://doi.org/10.1016/j.triboint.2015.10.012.
23. Ferrer, B. P., Biazon, L., Zuin, A., Tayer, P., Toro, A., & Cousseau, T. (2023). Rail grease formulation effect on its tribological performance under pure sliding conditions. *Tribology Transactions*, 66(1), 104–116. https://doi.org/10.1080/10402004.2022.2123754
24. Georgiou, E. P., Drees, D., De Bilde, M., Feltman, F., & Anderson, M. (2020). Grease adhesion and tackiness: Do they influence friction? *NLGI Spokesm*, 84, 20–25.
25. Chen, Y., Jha, S., Raut, A., Zhang, W., & Liang, H. (2020). Performance characteristics of lubricants in electric and hybrid vehicles: A review of current and future needs. *Frontiers in Mechanical Engineering*, 6. https://doi.org/10.3389/fmech.2020.571464.
26. Andrew, J. M. (2019). The future of lubricating greases in the electric vehicle era. *Tribology & Lubrication Technology*, 75, 38–44.
27. Farfan-Cabrera, L. I. (2019). Tribology of electric vehicles: A review of critical components, current state and future improvement trends. *Tribology International*, 138, 473–486. https://doi.org/10.1016/j.triboint.2019.06.029.
28. Sanchez Garrido, D., Leventini, S., & Martini, A. (2021). Effect of temperature and surface roughness on the tribological behavior of electric motor greases for hybrid bearing materials. *Lubricants*, 9(6), 59. https://doi.org/10.3390/lubricants9060059.
29. Abdul Razak, I. H., Ahmad, M. A., & Puasa, S. W. (2019). Tribological and physiochemical properties of greases for rail lubrication. *Tribology Online*, 14(5), 293–300. https://doi.org/10.2474/trol.14.293.
30. Biazon, L., Ferrer, B. P., Toro, A., & Cousseau, T. (2021). Correlations between rail grease formulation and friction, wear and RCF of a wheel/rail tribological pair. *Tribology International*, 153, 106566. https://doi.org/10.1016/j.triboint.2020.106566
31. Dhanola, A., & Garg, H. C. (2020). Tribological challenges and advancements in wind turbine bearings: A review. *Engineering Failure Analysis*, 118, 104885. https://doi.org/10.1016/j.engfailanal.2020.104885.
32. Peng, H., Zhang, H., Shangguan, L., & Fan, Y. (2022). Review of Tribological Failure Analysis and lubrication technology research of wind power bearings. *Polymers*, 14(15), 3041. https://doi.org/10.3390/polym14153041.
33. Georgiou, E. P., Drees, D., Lopes, L. M., De Bilde, M., & Koutsomichalis, A. (2023). Investigating the effect of temperature and time on the starvation of lubricants for the wind turbine industry. *Tribology - Materials, Surfaces & Interfaces*, 17(3), 179–186. https://doi.org/10.1080/17515831.2023.2232991.
34. Kücükpinar, E., & Langowski, H.-C. (2012). Adhesion aspects in packaging. *Journal of Adhesion Science and Technology*, 26(20–21), 2317–2324. https://doi.org/10.1163/156856111x610108.

35. Georgiou, E. P., Drees, D., Lopes, L. M., & Gerlach, C. (2022). Measuring the frictional behavior and adhesion of pet bottles. *Lubricants*, 10(9), 204. https://doi.org/10.3390/lubricants10090204.
36. Michalski, M., Desobry, S., & Hardy, J. (1997). Food materials adhesion: A review. *Critical Reviews in Food Science and Nutrition*, 37(7), 591–619. https://doi.org/10.1080/10408399709527791.
37. Noren, N. E., Scanlon, M. G., & Arntfield, S. D. (2019). Differentiating between tackiness and stickiness and their induction in foods. *Trends in Food Science & Technology*, 88, 290–301. https://doi.org/10.1016/j.tifs.2019.03.014.
38. Sonneveld, M. H., & Schifferstein, H. N. J. (2008). The tactual experience of objects. *Product Experience*, 41–67. https://doi.org/10.1016/b978-008045089-6.50005-8.
39. Kergoat, M., Giboreau, A., Nicod, H., Faye, P., Diaz, E., Beetschen, M., & Meyer, T. (2012). Consumer preference for tactile softness: A question of affect intensity? *Journal of Sensory Studies*, 27(4), 232–246. https://doi.org/10.1111/j.1745-459x.2012.00388.x.
40. Biswas, S., & Visell, Y. (2019). Emerging material technologies for haptics. *Advanced Materials Technologies*, 4(4). https://doi.org/10.1002/admt.201900042.

10 Grease Degradation/ Water Resistance

Lijesh Koottaparambil and Michael M. Khonsari

10.1 INTRODUCTION

Lubricating grease [1] has evolved over thousands of years from natural compounds used in axles and chariot wheels to advanced soap-thickened formulations used in components of modern industry. In the Egyptian and Roman eras, grease was made from animal fats, plant oils, wool, and resins to reduce friction in simple machines. The Industrial Revolution brought significant advancements in grease technology, with soap-thickened greases like sodium and calcium-based greases transforming lubrication practices in various industries. The 20th century saw the development of synthetic lubricants such as silicone and fluorinated greases to meet the demands of aerospace and automotive applications. Modern additives like antiwear agents and rust inhibitors have further improved grease performance. Recent progress in chemistry and engineering has led to the creation of specialty greases tailored to specific industries, with a focus on environmentally friendly formulations and bio-based lubricants to enhance efficiency and sustainability [2].

Modern greases are made of a thickening agent of 3%–30% suspended in a base oil (70%–90%) with the remaining percentage of suitable additives [1, 3–5]. Different concentrations of thickener in the base oil result in varying grease consistencies (refer to Figure 10.1), as classified according to the guidance by the National Lubricating Grease Institute (NLGI). The NLGI consistency grades range from 000 (very soft) to 6 (very stiff). The stiffness of grease results from the physical interaction between the thickener and lubricating oil, and the desired consistency requires using thickeners made of soap or non-soap or their hybrid. Soap-thickened grease is the most common

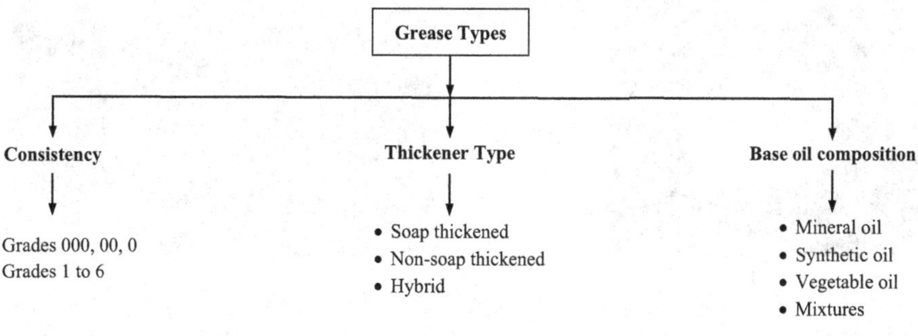

FIGURE 10.1 Classification of greases.

DOI: 10.1201/9781003496069-10

type of grease, where the thickener is typically a metallic soap—e.g., lithium, calcium, sodium, or aluminum—while non-soap thickened greases use alternative thickeners, such as polyurea, lithium complex, polymer, clay, or silica, instead of metallic soaps. The base oil can be mineral, synthetic, and vegetable oil-based greases. Greases that use mineral oils are derived from petroleum as the base oil, and greases formulated with synthetic base oils use polyalphaolefins (PAO), esters, poly-alkylene glycols (PAG), or silicones. Compared to mineral oils, synthetic base oils offer superior characteristics such as low pour points, high-temperature stability, and oxidative resistance [6]. However, they are infrequently employed in multipurpose greases due to their higher cost. Vegetable oil-based greases are typically made from soybean or rapeseed oil, renewable and environmentally friendly alternatives.

Compared to lubricating oils, grease possesses a high apparent and complex viscosity, offers good sealing properties, demands less frequent maintenance, does not require auxiliary supply arrangements and/or recirculation, corrosion protection, and can even tolerate operating in a contaminated environment [5,7]. The semi-solid nature allows it to adhere to surfaces and form a protective film for long-lasting lubrication under extreme conditions such as high temperatures, heavy loads, varying speeds, and corrosive environments. Thus, properly formulated grease can reduce friction and wear, extending the lifespan of machinery and equipment. Owing to these benefits, 90% of rolling element bearings are grease lubricated, which is also widely applied to various industrial components such as pin bushings, journal bearings, gears, slideways, and the like [8]. Nevertheless, the finite working life of grease is significantly impacted by its degradation, which is influenced by mechanical, thermal, and oxidative stability. While in some equipment—e.g., CV joints—the grease may last longer than the metal components, in general, the lifespan of grease is shorter than that of the rolling element bearings. It must be selected to protect the bearing surfaces and requires replenishment. Thus, it is crucial to understand the mechanisms of grease degradation and replenishment requirements to ensure machinery's continuous, reliable operation.

The primary cause of failure, accounting for approximately 55%, is related to the lubrication of rolling element bearings (see Figure 10.2) [9]. It is important to note that rolling element bearing failure is responsible for 51% of motor failures [10]. A failed grease, oxidized and hardened (as shown in Figure 10.3a [11]), renders

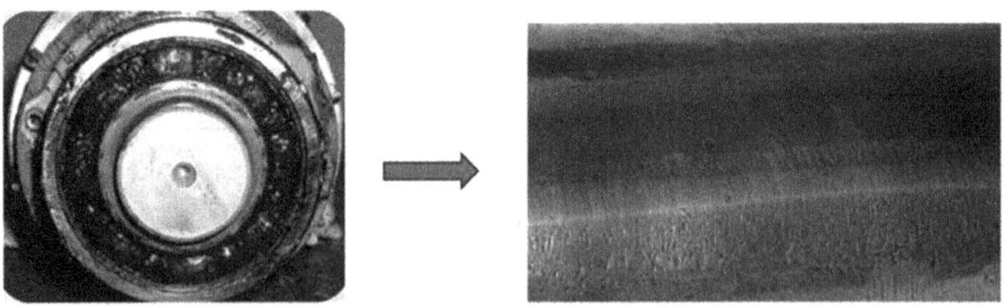

FIGURE 10.2 Grease and bearing failure. (a) Hardening of grease [11] and (b) worn surface of the bearing surface [12].

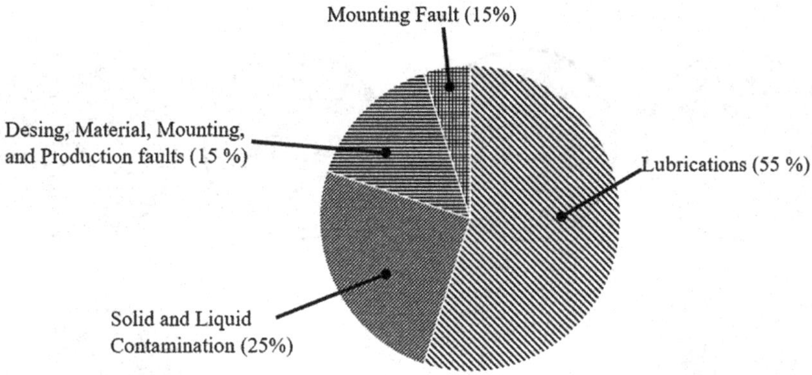

FIGURE 10.3 Causes of failure in rolling bearings [9].

the bearing inaccessible for re-lubrication, preventing the newly added grease from displacing the existing failed grease within the bearing. This obstruction to grease displacement leads to overheating, high friction, and wear in the roller contact area, resulting in bearing (see Figure 10.3b [12]). As per a report [13], hammering and chiseling old grease out of the bearings from the kiln cars and re-lubrication resulted in considerable expenditure annually. Despite frequent re-lubrication, typically, two kiln car-bearing failed each month, leading to considerable downtime [13]. Note that using the appropriate grease and re-lubrication scheduling can prevent bearing failure. Therefore, accurately determining the lifespan of grease is critically important for users and operators to properly schedule re-lubrication intervals and prevent catastrophic component failures.

Grease degradation is an irreversible process associated with permanent changes in its structure. It can be broadly categorized into physical/mechanical [14–17] and chemical degradation [18,19] mechanisms. In the following sections, each type of degradation of grease is discussed.

10.2 MECHANISMS OF GREASE DEGRADATION

10.2.1 Physical/Mechanical Degradation

Mechanical degradation, induced by shearing action, can profoundly impact the long-term performance of grease-lubricated ball and roller bearings. Grease that lacks mechanical stability may degrade in consistency under the mechanical forces exerted by the bearing's rolling elements. This degradation can result in grease leakage through seals or complete bearing failure in the worst-case scenario [20]. The effects of physical/mechanical degradation are significant and include the following phenomena [21]:

- Increased bleeding rate and oil leakage,
- Destruction of the thickener structure due to working and shearing,
- Contamination of the grease by foreign particles,
- Loss of volatile oxidation products at elevated temperatures.

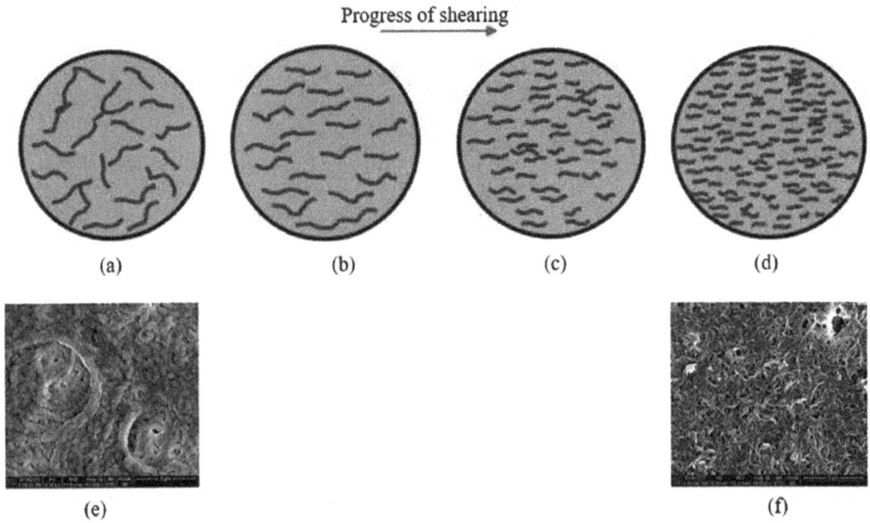

FIGURE 10.4 Physical degradation of the grease by the breakage of the thickener during the shearing: (a) $t_1 = 0$ s; (b) $t_1 < t_2 < t_3$; (c) $t_2 < t_3 < t_4$; (d) $t_4 > t_3$, [23]; (e) SEM image of thickeners pristine grease; (f) SEM image of thickeners after 15 h of shearing [24].

The shearing of grease occurs in two distinct phases: the churning phase and the bleeding phase [22]. The churning phase marks the initial stage of shearing following the filling of the bearing with grease, and it is distinguished by the observable macroscopic movement of the grease. Plenty of lubricant is available in this phase, and the contacts remain fully flooded. The initial randomly aligned thickener chains align in the direction of the shearing (see Figure 10.4a and b) and most of the grease ends up under the cage. In the subsequent bleeding phase, grease releases oil by phase separation, and the thickener chains break further with shearing (depicted in Figure 10.4c and d), reducing the grease viscosity [23]. Although the grease has a higher viscosity during the bleeding phase than its carrier oil, it no longer supplies sufficient lubricant to the contacts. The scanning electron microscope (SEM) images of pristine l grease thickeners and images of the same grease after 15 h of shearing at a shear rate of 5000 s shown in Figure 10.4e and f illustrate the breakage in the grease structure [24].

Lijesh and Khonsari [23] attempted to characterize the phase change and the grease degradation due to the thickener's breakage. They installed a torque sensor between the motor and grease-lubricated ball bearing and measured the change in the torque value with time. The measured change in the grease viscosity is indicative of its degradation. The measured torque at 1500 rpm shown in Figure 10.5 reveals two distinct regions: an initial transient region and a steady-state region with steady degradation. The transient region indicates the churning phase, where the thickeners align with the shearing, and the steady state degradation region represents the bleeding phase, where the thickeners break down and the viscosity of the grease steadily reduces [16,23,25,26]. Eventually, the viscosity decreases to a level where metal-to-metal contact ensues, eventually leading to bearing failure [27]. Therefore,

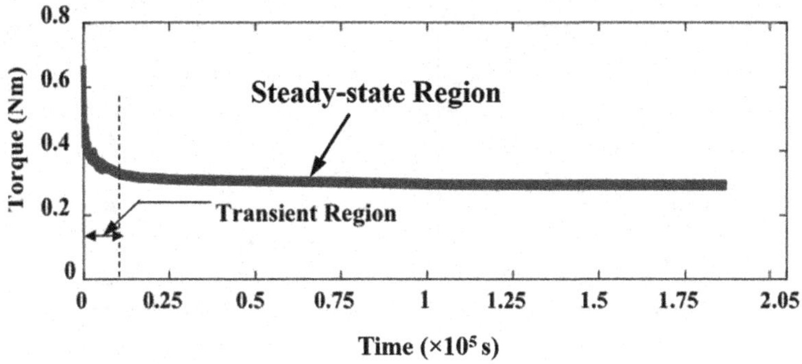

FIGURE 10.5 Torque value acquired from torque sensor for a grease-lubricated bearing [23]. Two distinct regions observed during the shearing of grease an initial transient region followed by a steady-state region.

grease must meet the necessary lubrication requirements for its intended application with appropriate properties in the steady-state region.

10.2.2 Thermal Degradation

The operating temperature in many electric motors and various industrial applications ranges from 40°C to 70°C [28], where mechanical degradation due to shearing alone is dominant. The grease service life significantly drops if the operating temperature exceeds 70°C, where 50% of the oil content of the grease is lost through bleeding with negligible evaporation. It is estimated that for an increase in temperature by 10°C–15°C, the service life of grease is reduced by a factor of 1.5–2 [28–30]. A detailed discussion on the drop in grease life due to temperature rise is provided by Khonsari and Booser [31,32]. In many applications, grease operates at high temperatures and experiences both mechanical and thermal degradation. However, mechanical degradation is the dominant degradation mechanism in the absence of any chemical reactions, such as oxidation.

Grease degradation at temperatures higher than room temperature—without any chemical reactions, such as oxidation—is attributed to thermal degradation. The temperature rise can stem from high shearing, operating conditions, or both. Thermal degradation can be viewed as a component of mechanical degradation, as both processes involve a change in oil viscosity and oil separation of the base oil, particularly the volatile shorter-chain paraffin [33–35]. These mechanisms are consistent between thermal and mechanical degradation. However, the severity of degradation is more pronounced during thermal degradation.

It should be noted that the grease performance at elevated temperatures depends on the grease thickeners and base oil used. Cyriac et al. [4] investigated the temperature dependency of yield stress and viscosity on greases developed using different thickener and base oils: (1) calcium sulfonate thickener with mixed mineral and synthetic oil (CaS/MS), (2) calcium sulfonate thickener with mixed mineral oil (CaS/S), (3) lithium thickener with mineral oil (Li/M), and (4) lithium complex thickener with polyalphaolefin oil (LiC/PAO). Their study revealed that the yield stress, determined

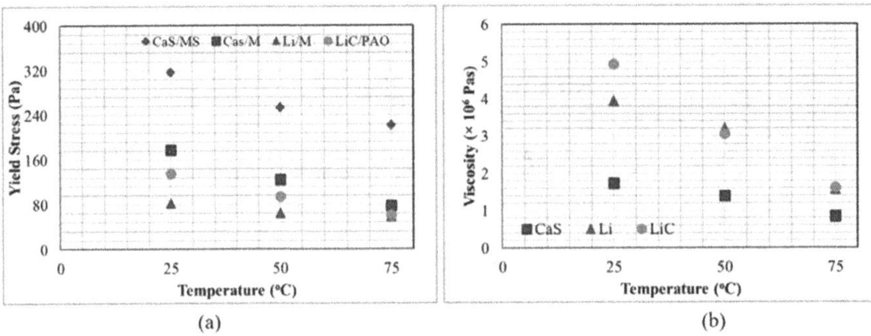

FIGURE 10.6 Yield stress and viscosity variation of calcium sulfonate (CaS), lithium (Li), and lithium complex (LiC) grease at temperatures: 25°C, 50°C, and 75°C: (a) yield stress and (b) viscosity [4].

from oscillatory measurements conducted at shear rates ranging from 10^{-8} to 100 s^{-1}, and the zero-shear-rate viscosity, derived from the Cross model, decreased by nearly half when the temperature increased by 50°C above room temperature.

Figure 10.6 also shows that grease performance at elevated temperatures depends on both the thickeners and the base oil types. Each grease has a specific temperature range within which its performance is optimal. Hence, selecting the appropriate grease for high-temperature applications requires careful consideration. Accordingly, SKF has developed a "traffic light concept" for the selection of grease types for a specific temperature range operation [22].

As defined by a traffic light analogy, the grease performance at different temperature ranges is divided into three color zones: green, yellow, and red zones. The optimum temperature for achieving a higher grease performance is defined in the 'green zone'. The green zone is limited by the low-temperature performance limit (LTPL) and the high-temperature performance limit (HTPL). The LTPL is established based on the temperature at which adequate oil bleeding persists [36], while the determination of the HTPL should rely on the point at which the grease life, assessed by methods like the R0F test rig [37], starts to deviate from the 'Arrhenius behavior'. The grease life decreases notably as the temperature drops below the LTPL until it reaches the low-temperature limit (LTL). At this level, the temperature is insufficient for the bearing to function without harm. This crucial temperature is determined through a start-up torque assessment. Conversely, the high-temperature limit (HTL) is defined by the grease's dropping point with an extra safety margin. The upper operating temperature can also be determined using high-temperature (HT) test rigs such as the rolling-object fatigue (ROF) rig or the four-ball extreme pressure 9 (FE9) test rig. In certain cases, other test rigs like the FE8 or the rolling-to-finish (R2F) may also be used for this purpose. Within the LTPL and HTPL window, the grease for the elevated temperature application can be selected.

10.3 CHEMICAL DEGRADATION

Chemical degradation involves a loss of antioxidants due to oxidation reactions rather than evaporation. Chemically degraded grease exhibits increased acidity following the depletion of antioxidants, and the formation of oxidized hydrocarbon species results in acidic and high-viscosity products, loss of the carboxyl bands in soap thickeners, and reactions with water to form acids or bases. These changes weaken its structure and diminish its lubricating efficacy. The details of the chemical degradation of grease from oxidation and water contamination perspective are discussed next.

10.3.1 Oxidation

One common form of chemical degradation is oxidation, which occurs when grease reacts with oxygen in the air and produces harmful by-products such as acids, peroxides, and lacquers. These by-products can corrode metal surfaces, degrade lubricating properties, and increase churning losses and friction, ultimately leading to equipment failure. Several factors contribute to grease oxidation, including elevated operating temperatures, the presence of metallic wear debris containing transition elements and their oxides, and the mechanical shearing of the grease that promotes the introduction of air into the lubricant. Mechanical shearing contributes to three to five percent promote oxidation. Ventilation of the bearing removes oil droplets [38] and help reduce the amount of free oil for replenishment, thereby increasing evaporation and oxidation [35]. Nevertheless, oxidation due to higher temperatures is the most influential factor in determining grease's service life. As pointed out in the previous section, the temperature range at which oxidation occurs varies depending on the specific grease thickener, base oil, and additive types.

It is important to note that existing literature provides varying temperature ranges for oxidation. For example, Hurley et al. [39] contend that within a temperature range of 40°C–100°C, a rise of 10°C effectively doubles the oxidation rate. Lugt [40] states that grease oxidation occurs above 120°C while Couronne and Vergne [41] observed no grease oxidation until 150°C. Therefore, it is necessary to test grease samples at different elevated temperatures to identify the temperature for starting oxidation. Generally, the Fourier Transform Infrared Spectroscopy (FTIR) test is performed on pristine and oxidized grease samples to gain valuable information about their chemical composition and structural changes over time. By comparing the FTIR spectra of pristine and oxidized grease, one can identify changes in functional groups, such as carbonyl (C=O), hydroxyl (OH), and sulfide (S–H) groups, which are indicative of oxidation and degradation processes. To understand the structure changes, it is necessary to understand the thermo-oxidation reactions prevailing in the grease at elevated temperatures.

10.3.2 Thermo-Oxidation Reaction

The degradation of the grease through thermo-oxidation follows the free radicals reaction mechanism [42], as illustrated in Equation (10.1). Here, RH denotes the base oil and thickener in grease, $R^\bullet$ and $H^\bullet$ are free radicals of the alkyl and hydrogen, respectively. Initially, alkyl free radicals are formed within the grease due to the rise in temperature during their shearing.

$$RH \rightarrow R^\bullet + H^\bullet \tag{10.1}$$

The alkyl free radicals react with oxygen gas (O_2) to quickly generate peroxide groups ($ROO^\bullet$).

$$R^\bullet + O_2 \rightarrow ROO^\bullet \quad (10.2)$$

Subsequently, the reaction between the peroxide groups and RH (representing the base oil and thickener) occurs, directly abstracting hydrogen atoms from RH. This process leads to the formation of secondary alkyl free radicals:

$$ROO^\bullet + RH \rightarrow ROOH + R^\bullet \quad (10.3)$$

The reactions would then continue until the chain termination, which could happen through the reaction with the hydrogen-free radical and/or mutual coupling, as illustrated in Equations (10.4) and (10.5), respectively.

$$R^\bullet + H^\bullet \rightarrow RH \quad (10.4)$$

$$R^\bullet + R^\bullet \rightarrow R-R \quad (10.5)$$

At elevated temperatures, ROOH is decomposed into RO• and HO•, as indicated in Equation (10.6). These groups then engage in further reactions with RH, leading to the production of the final alkyl radicals ($R^\bullet$), as shown in Equations (10.7) and (10.8). Subsequently, these radicals would undergo a further chain propagation process, contributing to the ongoing decrease in the thermal-oxidative stability of the lubricating grease.

$$ROOH \rightarrow RO^\bullet + OH^\bullet$$

$$OH^\bullet + RH \rightarrow H_2O + RO^\bullet$$

$$RO^\bullet + RH \rightarrow ROH + R^\bullet$$

As the chain process terminates, the concentration of oxidation products containing carbonyl groups, such as acidic compounds and high-viscosity substances, intensifies within the lubricating grease, significantly degrading its lubricating properties.

To elucidate the phenomenon of tribo-oxidation, the FTIR results comparing pristine and oxidized lithium-based grease containing MoS_2 particles, as obtained from [38], are presented in Figure 10.7a and b, respectively. In the spectrum of the pristine grease (Figure 10.7a), characteristic absorbance peaks are evident, including carboxylate stretch at 1597 cm and hydroxyl at 3442 cm, attributed to the presence of the hydroxy-stearate thickener. Additionally, bands at 1377 and 1460 cm are assigned to CH vibrations from the base oil [18]. The used grease's infrared spectrum (Figure 10.7b) shows a sharp and intense peak of the carbonyl (C=O) band at 1710 cm, indicating the oxidation of base oil and thickener within the grease. Consequently, it could be concluded that large amounts of compounds containing the C=O group were produced in the used greases. Furthermore, reducing the thickener peaks to a broader level suggests that the composition of the used grease sample primarily comprises base oil and carbonyl-containing degradation products. These observations confirm the thermo-oxidative degradation of the grease in their application. In this way, one can study the FTIR results and identify the presence of oxidation in the grease.

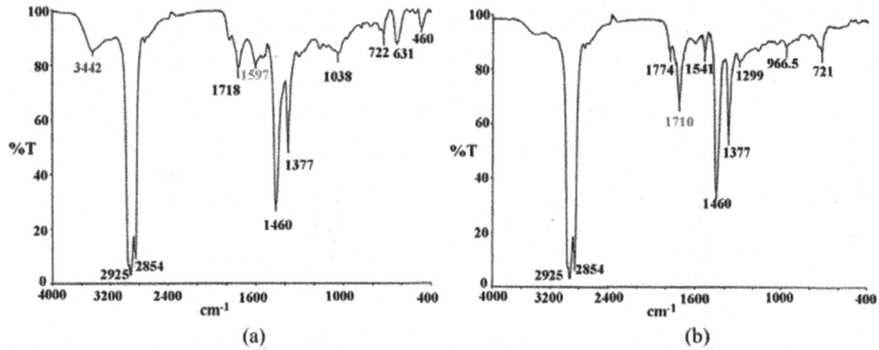

FIGURE 10.7 FTIR spectra of (a) pristine and (b) oxidized grease sample [42].

10.3.3 Water Contamination

Many studies indicate that the contamination of water significantly impacts the grease life. Water contamination tends to modify the grease's chemical composition and deteriorate its rheological characteristics. Consequently, this deterioration adversely impacts the functionality of machinery components. The importance of water contamination has been documented in numerous research reports, a selection of which is outlined below.

- According to Duncanson [43], water contamination accelerates the degradation of grease additives and promotes acid formation at a rate ten times faster.
- Water in bearings reduces film thickness and causes flash vaporization, erosive wear, and hydrogen embrittlement through hydrolysis [44].
- Even a mere 1% water content can diminish the lifespan of a journal bearing by up to 90% [45].
- Water levels as low as 0.02% (200 ppm) adversely affect the longevity of roller bearings [46].
- Increasing water content from 13% to 38% significantly enlarges the four-ball wear scar in calcium sulfonate complex grease from 0.38 to 0.48 mm [47].
- Emulsification of grease due to water leads to notable declines in adhesive and cohesive properties, particularly detrimental in steel rolling mills [48–50].
- A few greases harden in the presence of water, reducing their ability to lubricate contacts and resulting in heightened power consumption [47,48].
- In water-repellent greases, free water ingress through seals leads to corrosion and eventual bearing failure [49,51].
- A recent review article [52] underscores that many commercially available greases tout water resistance without quantitative substantiation, indicating the need for improved criteria to define grease water resistance properties.

Water contamination initiates various structural and chemical mechanisms that hasten grease degradation, underscoring the importance of selecting suitable

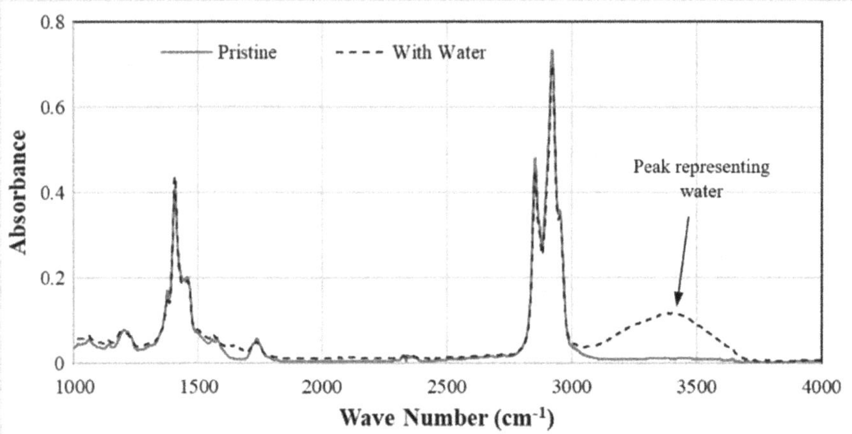

FIGURE 10.8 FTIR spectra of pristine and water-contaminated grease.

grease to extend the operational lifespan of critical machinery components. For instance, implementing a water-resistant grease in a pipe mill bearing reportedly decreased bearing failure due to water washout from four occurrences per month to just one, resulting in savings of $5500 in bearing expenses and $20,000 in production costs [53].

Like oxidated grease, water contamination in the grease can be determined by performing FTIR on the pristine grease and comparing the infrared spectrum with the water-contaminated grease. For example, the FTIR is performed on the pristine and water-contaminated CaS/m grease, and the results are provided in Figure 10.8. It shows an intense peak of the infrared spectrum at 3400 cm is observed, showing the presence of water in the grease [54].

Assessing the water resistance properties of grease by only measuring rheological properties can be misleading. Lijesh and Khonsari [55] observed that depending on the composition of the grease, the rheological properties of grease exhibit varied behavior in the presence of water. Some greases thicken when exposed to water, while others experience a thinning effect. Figure 10.9a–d show the log plots of shear stress and strain at 25°C for four types of grease: lithium complex in mineral oil (LiC/m) and calcium sulfonate in mineral oil (CaS/m), polyurea in mineral oil (PU/m), and white lithium grease in bio-degradable oil (WLi/b) greases. Figure 10.9a and b reveal that the shear stress values for pristine LiC/m and CaS/m greases are lower than water-contaminated greases, while Figure 10.9c and d suggest an opposite trend for PU/m and WLi/b greases. Therefore, this approach fails to assess the water-repellant properties in a single platform.

Lijesh and Khonsari [55] proposed an efficient technique to evaluate the water contamination in the grease by measuring the contact angle of a water droplet on the grease surface. A grease with a higher degree of contact angle exhibits minimal wettability and is hydrophobic. In contrast, grease that attracts water will display good wettability, appear hydrophilic, and show a smaller contact angle. Water-contaminated grease will have more wettability or hydrophilic tendency than

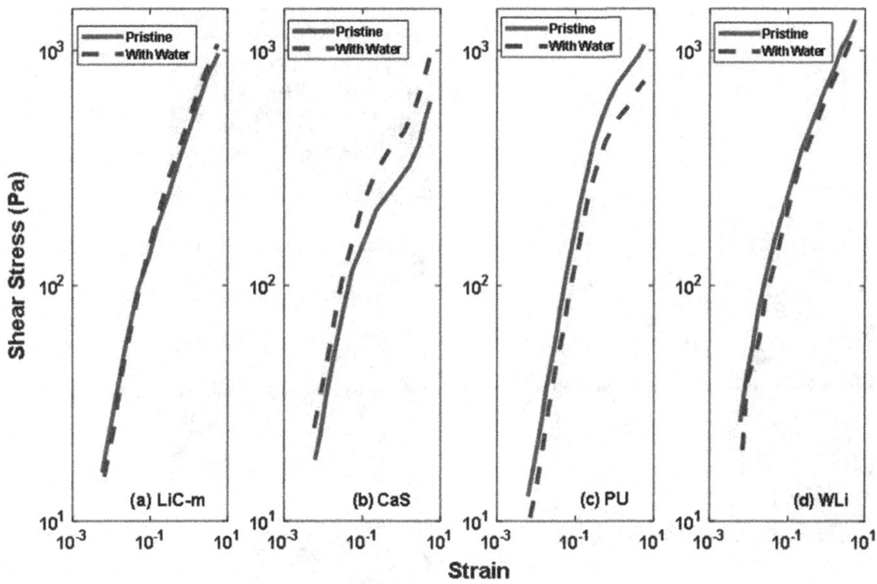

FIGURE 10.9 Log–log plot of shear stress versus strain for LiC/m, CaS/m, PU/m, and WLi/b greases [55].

pristine grease. Therefore, water-contaminated grease has a smaller contact angle compared to pristine grease.

The setup used for measuring the contact angle value of the water droplet on the grease surface is provided in Figure 10.10. This method further offers a quick tool for researchers seeking to improve hydrophobic or hydrophilic features while assessing different combinations of additive chemistries and thickeners. The contact angle value on the grease surface varies if the composition, thickener structure, and chemical affinity change. Therefore, this can be employed to assess the degradation level of grease due to shearing and oxidation.

10.4 GREASE FOR ELECTRICAL VEHICLE (EV)

Research on electric vehicles (EVs) is prominently focused on enhancing battery performance and developing energy management strategies. However, some components and systems that play a crucial role in vehicle efficiency and reliability are often overlooked, and lubricating grease is one such critical component [57]. In contemporary internal combustion engine (ICE) vehicles, greases are typically employed in around fifty locations. In contrast, there are up to 100 electrical motors in a typical electric EV, with an average count of 80 [58]. Given the significant number of electrical motors, selecting appropriate greases for these components is now more critical than ever.

Furthermore, the operating conditions of EV and IC engines differ significantly [59]. For instance, EV motors often operate at higher temperatures and higher speeds. Further, the leakage of electrical current and electromagnetic fields from electric

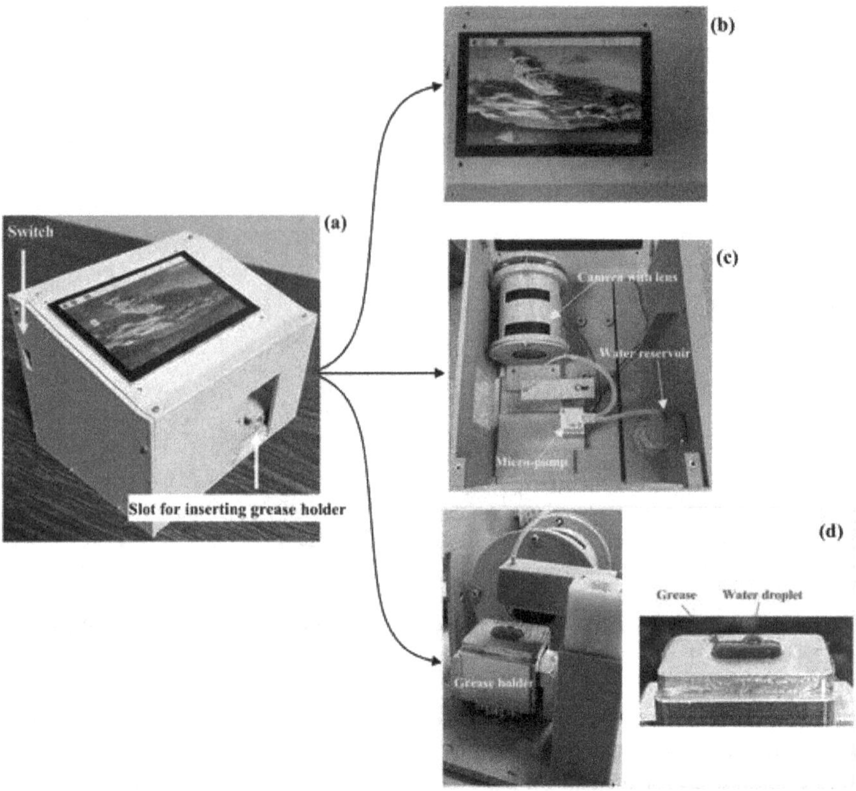

FIGURE 10.10 Custom-designed contact angle setup with their components: (a) isometric view of the contact angle setup; (b) display and microcontroller; (c) top view of the setup showing the camera and lens, micropump, and water reservoir; and (d) grease holder with grease and water droplet [56].

modules, sensors, and circuits generates arcing that tends to cause faster grease degradation. Therefore, it is essential to develop specialized greases tailored for specific applications in electric vehicles, ensuring safer and more reliable vehicle operation.

At lower speeds, the thickener plays a significant role in augmenting the film thickness. In comparison, at higher rotational speeds, the film formation is primarily influenced by the base oil component of the grease (see Figure 10.11 [60]). The increase in film thickness at higher speeds depends on the choice and formulation of the grease [61]. However, the shearing of grease during its application consumes energy. Considering that EVs contain nearly twice the amount of grease in electric motors compared to IC vehicles, one must carefully select appropriate to prioritize energy efficiency.

The efficiency of the motor drive of EVs, particularly at higher speeds, depends on the viscous force imparted by the grease components. Thus, low-viscosity greases are employed to enhance the energy efficiency of EVs and minimize friction at higher

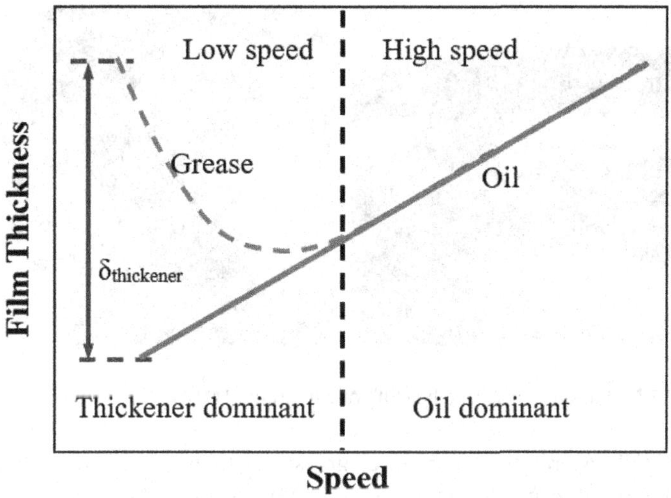

FIGURE 10.11 Schematic diagram showing the differences in film thickness between oil and grease, especially at low speeds [60].

rotational speeds. Generally, greases with synthetic oils are favored for their superior oxidation stability, and the choice of suitable thickeners for use in EV greases remains an ongoing area of research. Nevertheless, employing low-viscosity fluids leads to a reduction in the film thickness that tends to increase the operating temperature and reduce the bearing's fatigue life. Ideally, grease should generate films thick enough to separate mating surfaces fully but not excessively thick to cause energy loss during high-speed shearing.

Moreover, bearing components in EVs with the newly formulated greases with optimum grease viscosity will demand newer bearing geometries and surface topography to mitigate the effects of low-viscosity lubricants.

Along with long life and energy efficiency, the electrical conductivity of grease must be considered. If the grease's electrical conductivity is high, current-leaking/short circuits can occur. On the other hand, if conductivity is too low, it may cause a net charge buildup and result in electrically induced bearing damage [62]. Additionally, grease with low conductivity tends to oxidize and cause electrical arcing, ultimately reducing its capacity to safeguard the lubricated surface. Figure 10.12 shows a picture of a failed grease in a bearing due to arcing [11].

10.4.1 Grease Selection for EVs

Ensuring optimal performance and longevity of EV components necessitates a meticulous selection of grease composition, encompassing base oil type, thickener, and additives, and carefully considering their physical attributes, such as viscosity and viscosity index. This strategic approach mitigates the risk of grease failure, minimizes friction torque, and ultimately augments the durability of critical components within EV systems.

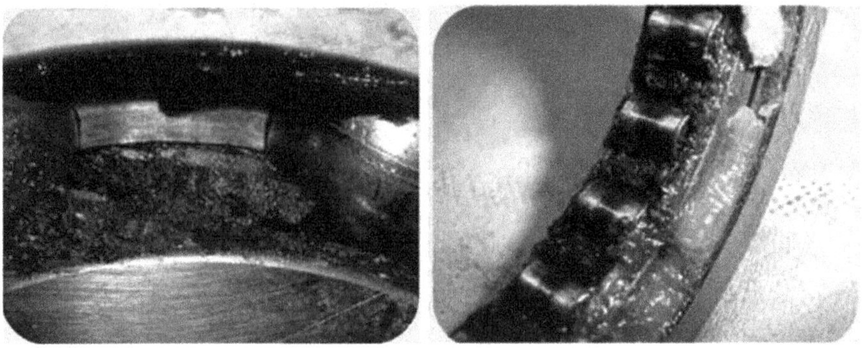

FIGURE 10.12 Failed grease in a bearing due to arcing [11].

- **Base Oil**: At room temperature, greases with low to medium viscosity yield 35%–40% in friction torque reduction, attributed to improved formation of lubrication film and decreased metal-to-metal contact. Conversely, excessively high viscosity grease led to a 50%–60% increase in frictional torque, resulting in elevated energy consumption [63,64]. Nevertheless, the thickener composition significantly influenced the grease performance. De Laurentis et al. [65] investigated the influence of various base oil types on the friction behavior of wheel bearings. They observed that synthetic oils exhibited lower friction coefficients and offered extended service life compared to mineral oils.

  Further, synthetic base oils offer enhanced thermal stability, low-temperature fluidity, and resistance to oxidation. Polyalphaolefins (PAOs), esters, alkylated naphthalenes, and poly-alkylene glycols (PAGs) are commonly used synthetic base oils in EV greases, providing excellent lubrication properties and extending component life. It is also essential to evaluate the compatibility of the base oil with seals and plastics used in EV components. Certain base oils may degrade rubber seals or interact adversely with plastic materials, leading to premature component failure. Compatibility testing should be conducted to ensure the selected base oil is suitable for use in EV applications. Finally, the overall conductivity of the grease can be altered by employing the synthetic base oils, such as polyalphaolefins (PAOs) or poly-alkylene glycols (PAGs), which exhibit higher conductivity compared to mineral oils due to their molecular structure.
- **Thickeners**: In high-temperature applications, the primary factor influencing friction torque is the type of thickener and base oils used. Greases with diurea/polyurea thickeners are preferred over lithium complex greases in EV applications due to their superior mechanical stability, high-temperature resistance, and compatibility with various additives [66]. They offer superior protection against wear and corrosion, making them ideal for the demanding conditions encountered in EV drivetrains. However, now industries are moving towards greases with polyurea thickeners, especially in applications requiring high-speed performance [67]. They provide excellent

shear stability along with compatibility with certain electrical components, making them suitable for specific EV applications. Sanchez Garrido et al. [68] investigated the impact of lithium-based, calcium-based, and polyurea-based greases on friction and wear performance using a ball-on-disc setup; these greases were formulated specifically for electric motor bearing applications. Their results revealed that calcium and polyurea-based greases exhibited reduced friction torque and enhanced wear resistance compared to lithium-based grease.

- **Additives**: At the high loads and pressures experienced by components in EV drivetrains, antiwear additives such as zinc dialkyl dithiophosphate (ZDDP) compounds are essential for protecting critical surfaces from wear and extending component life. On the other hand, organic moly compounds reduce friction but can also cause higher wear. Further, EVs often operate at high temperatures, particularly in battery compartments and electric motors, where antioxidants and thermal stability additives are required to prevent grease oxidation and thermal degradation. The antioxidant additives such as diphenyl amines of phenyl alpha naphthylamines, ZDDP, phosphite, and sulfur-based antioxidants ensure long-term performance, operational durability, and reliability. EVs are also exposed to various environmental conditions, including moisture and salt, which can accelerate corrosion. Corrosion inhibitors such as rust inhibitors and anticorrosion additives help protect metal surfaces from corrosion, preserving the integrity of components. Certain additives incorporated into grease formulations can affect conductivity. For example, graphite or metallic particles added for enhanced lubrication properties may increase conductivity, while insulating additives like ceramics or polymers may decrease conductivity.

10.5 CONCLUSIONS

Grease degradation encompasses physical and chemical processes, significantly impacting its performance and longevity. Physical degradation includes the degradation due to shearing, temperature fluctuations below oxidation, and high pressure—which leads to changes in breakage of the grease structure and consistency—ultimately affecting the lubricating ability of the grease. Chemical degradation includes oxidation and degradation due to water contamination, compromising the grease's effectiveness in providing adequate lubrication and protection.

Energy efficiency, reliability, and component longevity are paramount for EVs, and selecting the suitable grease is crucial. Grease formulations tailored for EV applications must withstand the unique operating conditions encountered in electric motors, including high speeds, heavy loads, current leakage, and temperature variations. In addition to electric motors, greases used in wheel bearings, steering systems, and CV joints must also meet these demands. For example, wheel bearings in EVs require greases that can handle high radial loads and maintain low friction under continuous operation. Similarly, steering systems and CV joints require greases with superior oxidation resistance, thermal stability, and compatibility with electrical components to ensure optimal performance and durability across all components.

REFERENCES

1. Raj Shah, W.T., *NLGI Lubricating Grease Guide,* 7th Edition, NLGI, Liberty, MO, US, 2022.
2. Syahir, A., et al., A review on bio-based lubricants and their applications. *Journal of Cleaner Production*, 2017. 168: p. 997–1016.
3. Tanguy, P.A., et al., Mixing performance induced by coaxial flat blade-helical ribbon impellers rotating at different speeds. *Chemical Engineering Science*, 1997. 52(11): p. 1733–1741.
4. Cyriac, F., P. Lugt, and R. Bosman, Impact of water on the rheology of lubricating greases. *Tribology Transactions*, 2016. 59(4): p. 679–689.
5. Lijesh, K. and M. Khonsari, On the assessment of mechanical degradation of grease using entropy generation rate. *Tribology Letters*, 2019. 67: p. 1–13.
6. Donahue, C.J., Lubricating grease: A chemical primer. *Journal of Chemical Education*, 2006. 83(6): p. 862.
7. Corporation, F.B., *Rolling Bearing Lubrication*, Vol. WL 81 115/4 EA, SKF, Gothenburg, Sweden, 2002.
8. Lugt, P.M., *Grease Lubrication in Rolling Bearings*, John Wiley & Sons, New York, 2012.
9. Radu, C., The most common causes of bearing failure and the importance of bearing lubrication. *RKB Technical Review*, 2010: p. 1–7.
10. Butche, J., *What to Do about Common Motor Bearing Failures*, Energy Management Corporation, Salt Lake City, UT, 2022.
11. Earthring, *Destructive Implications of Shaft Currents on Motor Bearings.* https://earthrings.co.za/shaft-currents
12. *Lubricant Failure = Bearing Failure, Machinery Lubrication.* https://www.machinerylubrication.com/Read/1863/lubricant-failure
13. BearingNews, Failure of kiln car bearings caused by high temperatures hardening the grease. *BearingNews*, 2016, https://www.bearing-news.com/failure-kiln-car-bearings-caused-high-temperatures-hardening-grease/.
14. Rezasoltani, A. and M. Khonsari, Mechanical degradation of lubricating grease in an EHL line contact. *Tribology International*, 2017. 109: p. 541–551.
15. Rezasoltani, A. and M. Khonsari, An engineering model to estimate consistency reduction of lubricating grease subjected to mechanical degradation under shear. *Tribology International*, 2016. 103: p. 465–474.
16. Kuhn, E., Tribological stress of lubricating greases in the light of system entropy. *Lubricants*, 2016. 4(4): p. 37.
17. Kuhn, E., Correlation between system entropy and structural changes in lubricating grease. *Lubricants*, 2015. 3(2): p. 332–345.
18. Cann, P., Grease degradation in a bearing simulation device. *Tribology International*, 2006. 39(12): p. 1698–1706.
19. Aranzabe, A., et al., Comparing different analytical techniques to monitor lubricating grease degradation. In: *NLGI Spokesman-Including NLGI Annual Meeting*, National Lubricating Grease Institute, Kansas City, Mo, 2006.
20. Lundberg, J. and E. Höglund, A new method for determining the mechanical stability of lubricating greases. *Tribology International*, 2000. 33(3–4): p. 217–223.
21. Gurt, A. and M. Khonsari, The use of entropy in modeling the mechanical degradation of grease. *Lubricants*, 2019. 7(10): p. 82.
22. Lugt, P.M., Modern advancements in lubricating grease technology. *Tribology International*, 2016. 97: p. 467–477.
23. Lijesh, K. and M. Khonsari, Procedure for evaluating the mechanical degradation of a grease. *NLGI Spokesman*, 2020. 84(1): p. 24–35.

24. Rezasoltani, A. and M. Khonsari, On monitoring physical and chemical degradation and life estimation models for lubricating greases. *Lubricants*, 2016. 4(3): p. 34.
25. Moore, R. and A. Cravath, Mechanical breakdown of soap-base greases. *Industrial & Engineering Chemistry*, 1951. 43(12): p. 2892–2897.
26. Kuhn, E., Description of the energy level of tribologically stressed greases. *Wear*, 1995. 188(1–2): p. 138–141.
27. Hoshino, M., Theory of grease lubrication: Technological trends of grease lubrication. *Japanese Journal of Tribology*, 2002. 47(1): p. 13–23.
28. Booser, R. and M. Khonsari, Grease life in ball bearings: The effect of temperatures. *Tribology & Lubrication Technology*, 2010. 66(10): p. 36–44.
29. Booser, E. and M. Khonsari, Grease and grease life. In: Y. W. Chung and Q. J. Wang (Eds.), *Encyclopedia of Tribology*, Springer, New York, NY, 2013.
30. Farcas, F. and M. Gafitanu, Some influence parameters on greases lubricated rolling contacts service life. *Wear*, 1999. 225: p. 1004–1010.
31. Khonsari, M.M. and E.R. Booser, An engineering guide for bearing selection. *Tribology & Lubrication Technology*, 2004. 60(2): p. 26–32.
32. Khonsari, M.M. and E.R. Booser, *Applied Tribology: Bearing Design and Lubrication*, John Wiley & Sons, New York, 2017.
33. McClintock, R., A laboratory study of wheel bearing grease high-temperature life. *NLGI Spokesman*, 1980. 43: p. 414.
34. Ito, H., M. Tomaru, and T. Suzuki, Physical and chemical aspects of grease deterioration in sealed ball bearings. *Lubrication Engineering*, 1988. 44(10): p. 872–879.
35. Hurley, S., P. Cann, and H. Spikes, Lubrication and reflow properties of thermally aged greases. *Tribology Transactions*, 2000. 43(2): p. 221–228.
36. Kühl, R., Ölabgabeverhalten bei Tiefen und Hohen Temperaturen-Einfluss Aufden Temperatureinsatzbereich eines Schmierfettes in Wältzlagern. *GfT Tribologie-Fachtagung*: p. 1–6.
37. Lugt, P.M., et al., The ROF+ methodology for grease life testing. *NLGI Spokesman*, 2013. 77(1): p. 18–27.
38. Larsson, P.-O., B. Jacobson, and E. Höglund, Oil drops leaving an EHD contact. *Wear*, 1994. 179(1–2): p. 23–28.
39. Hurley, S., P. Cann, and H. Spikes, Thermal degradation of greases and the effect on lubrication performance. In: D. Dowson, C. Taylor, and T. Childs (Eds.), *Tribology Series*, Elsevier, Amsterdam, 1998, pp. 75–83.
40. Lugt, P.M., A review on grease lubrication in rolling bearings. *Tribology Transactions*, 2009. 52(4): p. 470–480.
41. Couronne, I. and P. Vergne, Rheological behavior of greases: Part II—effect of thermal aging, correlation with physico-chemical changes. *Tribology Transactions*, 2000. 43(4): p. 788–794.
42. Yu, Z.-Q. and Z.-G. Yang, Fatigue failure analysis of a grease-lubricated roller bearing from an electric motor. *Journal of Failure Analysis and Prevention*, 2011. 11: p. 158–166.
43. Duncanson, M. Detecting and controlling water in oil, 2005, https://www.machinerylubrication.com/Read/787/detecting-water-in-oil.
44. Dixon, M., *Trouble the Water*, University of Alabama Press, Alabama, 1989.
45. Corporation, N. *Water In Oil Contamination*. 2001. https://www.machinerylubrication.com/Read/192/water-contaminant-oil.
46. Dittes, N., P. Marklund, and A. Pettersson. Mixing grease with water. In: *European Conference on Improvement in Bearing Technology*, Nieuwegein, the Netherlands, 23 February 2015 to 24 February 2015.
47. Li, J., et al., The typical application of calcium sulfonate complex greases in steel mills. *NLGI Spokesm*, 2018. 82: p. 32–38.

48. Nagarkoti, B., Water resistance property of greases: An outlook. In: *Proceedings of the NLGI-India 12th Lubricating Grease Conference*, Panaji, India, 2010.
49. Mistry, A. Performance of lubricating greases in the presence of water. In: *NLGI Spokesman-Including NLGI Annual Meeting*, National Lubricating Grease Institute, Kansas City, Mo, 2005.
50. Leckner, J. Water+ grease = fatal attraction. In: *25th ELGI Annual General Meeting*, Amsterdam, the Netherlands, 2013.
51. Cann, P., Starved grease lubrication of rolling contacts. *Tribology Transactions*, 1999. 42(4): p. 867–873.
52. Gurt, A. and M. Khonsari, An overview of grease water resistance. *Lubricants*, 2020. 8(9): p. 86.
53. Coe, C., Grease: A Niche market opportunity. *Tribology & Lubrication Technology*, 2015. **71**(5): p. 32–34.
54. Wright, J., *Benefits of FTIR Oil Analysis*, 2015. https://www.machinerylubrication.com/Read/30205/ftir-oil-analysis.
55. Lijesh, K., M. Khonsari, and R.A. Miller, Assessment of water contamination on grease using the contact angle approach. *Tribology Letters*, 2020. 68: p. 1–12.
56. Lijesh, K., et al., The standard for assessing water resistance properties of lubricating grease using contact angle measurements. *Lubricants*, 2023. 11(10): p. 440.
57. Chen, Y., et al., Performance characteristics of lubricants in electric and hybrid vehicles: A review of current and future needs. *Frontiers in Mechanical Engineering*, 2020. 6: p. 571464.
58. Fish, G. Using the right greases for electric vehicles, 2021. https://www.fleetequipmentmag.com/greases-electric-vehicles/.
59. Peskoe-Yang, L., Electric vehicles make grease's future uncertain. *Tribology & Lubrication Technology*, 2020. **76**(2): p. 18–20.
60. Vengudusamy, B., C. Enekes, and R. Spallek, On the film forming and friction behaviour of greases in rolling/sliding contacts. *Tribology International*, 2019. 129: p. 323–337.
61. Cyriac, F., et al., Effect of thickener particle geometry and concentration on the grease EHL film thickness at medium speeds. *Tribology letters*, 2016. 61: p. 1–13.
62. Shah, R., et al., Grease performance requirements and future perspectives for electric and hybrid vehicle applications. *Lubricants*, 2021. 9(4): p. 40.
63. Kanazawa, Y., R.S. Sayles, and A. Kadiric, Film formation and friction in grease lubricated rolling-sliding non-conformal contacts. *Tribology International*, 2017. 109: p. 505–518.
64. Jain, A., et al., Impact of grease physical properties on the friction torque and service life of wheel bearings. *Proceedings of the Institution of Mechanical Engineers, Part J: Journal of Engineering Tribology*, 2024: p. 13506501241240020.
65. De Laurentis, N., et al., The influence of base oil properties on the friction behaviour of lithium greases in rolling/sliding concentrated contacts. *Tribology Letters*, 2017. 65: p. 1–16.
66. Root, J.C., A comparative study of polyurea and lithium complex grease thickeners. *NLGI Spokesman*, 1994. 58(9): p. 22–24.
67. Andrew, J.M., The future of lubricating greases in the electric vehicle era. *Tribology & Lubrication Technology*, 2019. **75**(5): p. 36–40.
68. Sanchez Garrido, D., S. Leventini, and A. Martini, Effect of temperature and surface roughness on the tribological behavior of electric motor greases for hybrid bearing materials. *Lubricants*, 2021. 9(6): p. 59.

11 Electrified Pin-on-Disk Testing of Greases

Jack R. Janik, Robert L. Jackson, and Sudip Saha

11.1 INTRODUCTION

11.1.1 JUSTIFICATION FOR ELECTRIFIED TESTING

The concern for equipment reliability in electrified conditions, such as wind turbines for power generation, has been a standing concern [1]. This concern has gained increased interest recently, particularly with the growing focus on electric vehicles (EVs) as alternative transportation. Like wind turbines, the electrified conditions within EVs affect internal powertrain components and can impact safety, efficiency, and reliability. While adding an electric potential affects EV-bearing performance, loading conditions significantly differ from those in internal combustion engine (ICE) vehicles. Compared to ICE systems, EV lubrication regimes along a Stribeck Curve transition much quicker from more demanding torque characteristics. Electric motors deliver immediate power for higher acceleration across comprehensive speed ranges, suggesting a faster and potentially more damaging boundary lubrication regime Likewise, grease choice directly impacts the friction generated within mechanical components, dramatically affecting the efficiency and range of EVs [2]. For EVs, lubricants with improved heat transfer can provide increased resistance to degradation from thermal loading. The advantageous properties of greases make them suitable replacements for traditional lubricants for applications requiring increased torque and load support [3].

As it pertains to wheel and shaft bearings for electric motors, 40%–60% of premature electric motor failures occur due to lubrication starvation or unoptimized [4–6]. 80%–90% of rolling element bearings are grease lubricated due to its superior load support and frictional reduction [7,8]. As modern EVs increase in complexity, additional components will demand specialized lubrication. Despite multiple opportunities for grease optimization in EVs, the functionality of grease is challenged in sealed (maintenance-free) bearings intended for extreme condition applications and adjustable speed drives [6]. Nonetheless, adjusting grease compatibility in electrified systems is an encouraging prospect for mitigating electrical bearing damage.

While standardized testing for ICE lubricants has been refined over decades, EV-specific fluids are relatively new and often depend on original equipment manufacturers (OEMs). As a result, the development of standardization efforts for EV fluids

DOI: 10.1201/9781003496069-11

is still in its early stages. In the meantime, it is crucial to develop tests that simulate realistic operating conditions. Currently, ASTM G99-23 offers a standard method for testing wear resistance and friction properties in a pin-on-disk or ball-on-disk application [9]. While this method provides valuable insight into sliding tribological systems, it neglects the effects of rolling. Likewise, ASTM G182-13 provides a simple way of determining breakaway friction characteristics of rolling element bearings, but it does not reflect the unique influence of individual rolling elements in a system [10]. Future iterations of these tests should incorporate an electrical potential across the lubricated contact to fully simulate EV damage mechanisms (e.g., electrical pitting and corrosion), thermal heat generation and dissipation, material compatibility with both conductors and semiconductors, lubrication efficiency, reliability, and service life.

11.1.2 Causes and Effects of Electrified Bearing Damage

To fully grasp the severity of electrically induced bearing damage (EIBD), it is essential to understand how electric currents are generated. Bearings can be exposed to instantaneous or sustained electric currents flowing across lubricated surfaces in electric motors [4,11]. These currents are generated by an inverter which converts the DC supply to an AC power supply at variable outputs and or by shaft voltages created by asymmetric magnetic pole distributions or manufacturing defects [12–17]. Often, these currents make their way to the bearing, where the relativity-high dielectric strength of the lubricant acts as a resistor-capacitor (RC) system. A relatively non-conductive lubricant can facilitate charge accumulation, eventually leading to arcing or discharging across the lubricating film [18–22]. The discharging process can cause surface damage and accelerate failure due to noise, vibration, and premature corrosion [4]. With the increased quantity and complexity of electrical systems in EVs, unintended electrical conduction of leakage currents through powertrain systems is inevitable. Nonetheless, studying and understanding the performance of mechanical components in electrified systems will support the sustainable development of electric transportation.

Recognizing the common damage patterns experienced by bearings affected by EIBD is also essential. In addition to traditional bearing failures from fatigue fractures (spalling), thermal cracking, smearing, skidding, and abrasive wear, exposure to shaft voltages and bearing currents accrue additional morphological failures [23]. At the highest magnification, *pitting* persists as a deep crater from a single, high-voltage discharge at low rotational speeds [11]. As current is discharged across the rolling contact, the discharge ablates the metal. Like that shown in Figure 11.1, pits are generally circular and contain high concentrations of carbon and oxygen, indicating that lubricant degradation and surface damage co-occur [24]. As pits increase with prolonged exposure to electrified conditions, individual craters merge, and damage progresses in a satin-like surface finish [25]. Satin-like changes to the surface finish are seen in Figure 11.2. These visible changes on running surfaces justify the term *frosting* and are visible to the naked eye or under a benchtop microscope. While other forms of EIBD exist, pin-on-disk tests are most

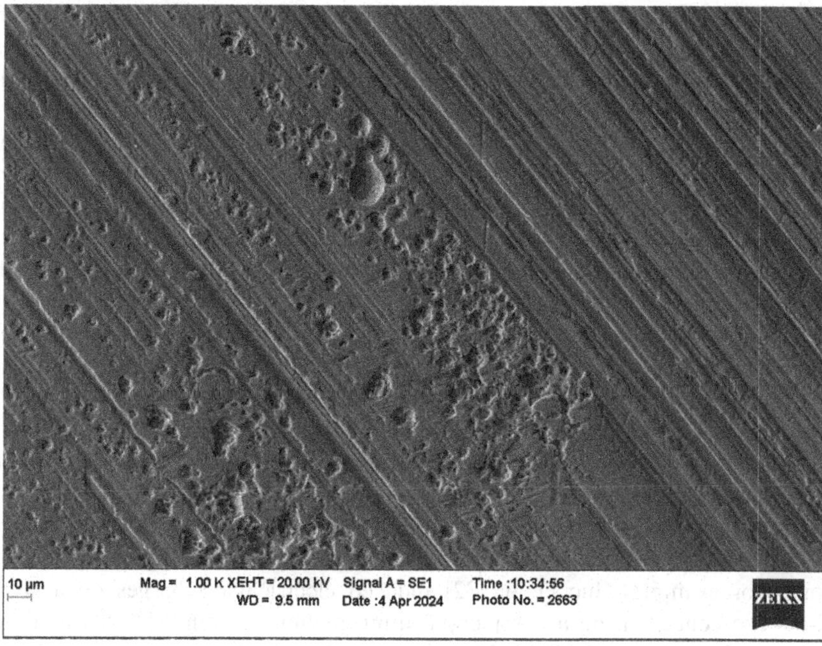

FIGURE 11.1 Electrically induced pitting using mineral base grease (mineral base oil + polyurea thickener) reciprocating across 2 cm tracks tested for 5400 seconds.

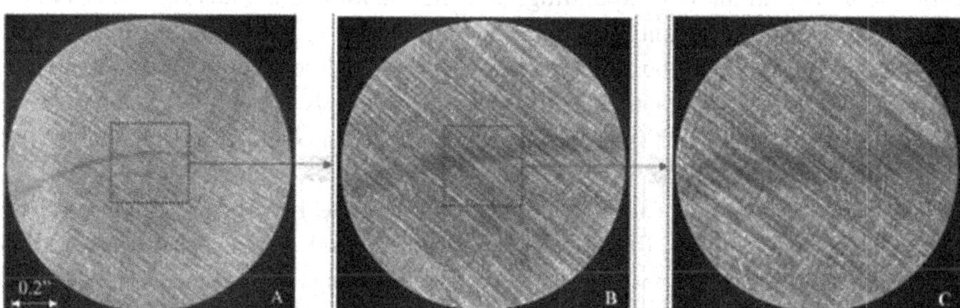

FIGURE 11.2 Frosting from pit accumulation. (Conditions were identical to those outlined in Figure 11.1 modified from [26].)

susceptible to these forms of damage. Figures 11.1 and 11.2 are the results of tests introduced later in this chapter.

11.1.3 Introduction to Electrified Tribology Testing

As machines, vehicles, and systems become increasingly electrified, more scientists and researchers are investigating the causes and effects of this electrification. Likewise, global work is being conducted to analyze bearing performance in electrified equipment [25–30]. Most studies conducted by these groups explored the behavior of complete rolling element bearings under electrified conditions. These tests were often performed for days or weeks before assessing for any damage. While using complete bearings closely mimics EV hardware and operating conditions, it does not promote a fundamental understanding of the arcing phenomena. Complete bearings operating under rolling conditions have numerous contact points, making it difficult to pinpoint the specific flow of electricity. During rotation, the individual loading of rolling elements is transient, causing localized changes in film thickness. With a dielectric lubricant, these changes in film thickness dramatically affect the location and severity of electrical discharging.

To simplify the complicated multi-body dynamics of complete bearings, electrified testing of single-bearing elements or mechanical components may shed light on EIBD formation. Demas et al. employed a modified Micro Pitting Rig (MPR) to compare electrical effects on the formation of white etching cracks (WECs), a common failure mode for wind turbine rolling element bearings [31]. While this test does not use a bearing element, it still represents a simplified method of understanding the effect of varied electrical currents on EIBD, particularly WEC-induced pitting.

Other work, however, has incorporated a single element under electrified conditions. For example, Guo et al. [32] induced electrical discharges on a single ball-on-disc contact using a commercial mini-traction machine (MTM) modified for electrical applications. MTM rigs are discussed in more detail in another chapter. Additionally, Tajedini et al. [33] conducted sliding ball-on-disk tests exploring the dynamic tribo-electrical behavior of nanomaterial-enriched lubricants. Research in this field is continuously expanding, and the mentioned studies represent only a selection of the most relevant research efforts. In single-element studies, test parameters are chosen such that the single rolling element are loaded comparably to an element within a full vehicle bearing. This load varies widely depending on the bearing size and speed rating, but it is not uncommon for the maximum pressure in rolling element bearings to exceed 3–4 GPa [34].

While the behavior of rolling elements in complete bearings is dominated by rolling rather than sliding, these tests serve as a crucial starting point for analyzing electrical and tribological phenomena. By focusing on a single contact area, researchers can more effectively manage and monitor the electric current through the test components and observe how bearing damage varies under different lubrication regimes and motions. This chapter will present a new test rig incorporating rolling ball-on-disk contacts to better understand EIBD under electrified rolling conditions.

11.2 TESTING EQUIPMENT AND INSTRUMENTS

The Bruker UMT Tribolab, renowned for its precision and versatility, is the premier choice for conducting tribological tests on electrified contacts. Utilizing a power screw loading mechanism, the machine applies a constant vertical load, while a load cell mounted to a leaf spring suspension compensates for irregularities, ensuring continuous load application throughout the test cycle. The horizontal movement of the sample is achieved via a slider mechanism driven by a power screw. The lower drive features a belt-driven spindle that rotates at user-defined speeds, providing precise control over test conditions. Despite its versatility, the Tribolab may require modifications or custom tooling to enable safe and efficient electrified testing.

When electrifying a tribological interface, the power and orientation of the electrical circuit can vary significantly. A key consideration is the small contact area typically found in pin-on-disk contacts. Excessive supply potential and current can lead to component overheating and electrical hazards. Applying resistance in series with the appropriate power supply lead is essential to mitigate this, effectively limiting current flow. Heating elements are a popular choice for their cost-effectiveness and long lifespan. Alternatively, incandescent or LED light bulbs can achieve similar performance. Regardless of the resistive element chosen, their power rating and configuration within the electrical circuit enable precise and accurate control of the interface power at the tribological contact.

Reliable delivery of potential and current through tribological components depends on mechanical loading conditions. The movement profile (e.g., reciprocating, complete rotations) significantly influences the optimal current delivery method. Depending on the radial displacement in each test cycle, specific current delivery methods may be more straightforward and more effective. Fully rotating tests offer more widespread wear patterns, stable film generation, enhanced heat dissipation, and simplified analysis. However, these tests prevent bolted or soldered connections, as wires or leads would inevitably become dangerously tangled. Consequently, as shown in Figure 11.3, a spring-loaded contact is required to maintain continuous electrical conduction throughout the test duration. While this variant solves the issue of full rotation, spring-loaded contacts introduce additional noise that may distort the signal from the tested contact, complicating the identification of discharge signals. Additionally, these contacts may wear out and require periodic replacement. Instead of a spring-loaded contact, slip rings

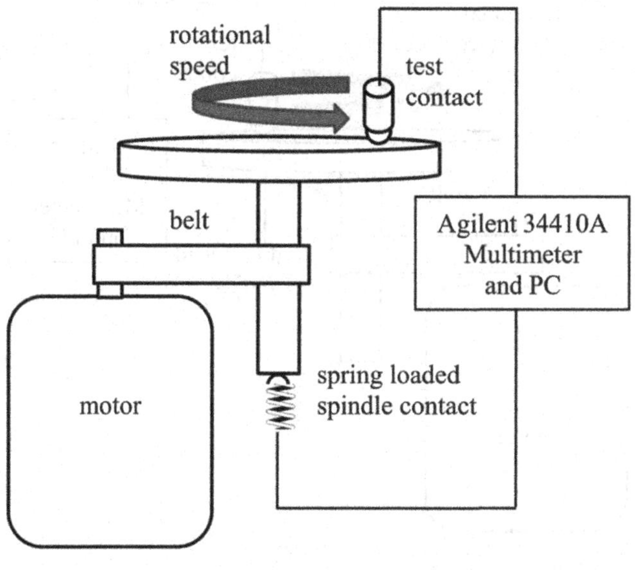

FIGURE 11.3 Schematic of a fully rotating electrical pin-on-disk test rig used in electrical connector lubricant evaluation. (Used with permission from [35].)

or brushes could also be used. However, these contacts also require periodic replacement and may contribute to some abrasive wear on mechanical components.

While fully rotational testing has advantages, reciprocating tests offer significant value. Many real-world applications, such as pistons, linear bearings, sliding seals, or bearings under start/stop conditions, are better represented by reciprocating tests, which effectively showcase localized wear. The contact area for reciprocating tracks is smaller than fully rotational tests, allowing for more consolidated and repeatable measurements. When examining end of test samples for morphologic damage, reciprocating tests reduced the surface area where damage is expected and made small-scale damage easier to find with visual techniques, such as SEM. While abrasive and adhesive wear along the confined contact path may affect contact area, electrically induced wear patterns differ significantly from mechanical wear and typically result in crater-like formations rather than material flattening or smoothening. Additionally, the shorter stroke length of reciprocating tests enables the use of bolted contacts, as shown in Figures 11.4 and 11.5, which reduces signal noise. It is also important to note that these tests do not involve a high-power supply, as the wires and contacts used are generally not rated for high-amperage applications.

The electrical circuit's fundamental configuration may remain unchanged whether spring-loaded or rolling contacts are employed. For this application, where the electrical supply and the desired measured signal across a contact are connected in parallel, a 4-wire (Kelvin) test method is recommended. This approach is advantageous for testing electrical contacts as it effectively eliminates the influence of resistances in the lead and contacts, providing more accurate measurements of the contact's inherent resistance. The idealized circuit for this application is illustrated in Figure 11.6. As depicted in Figure 11.6, the two parallel circuits converge at the

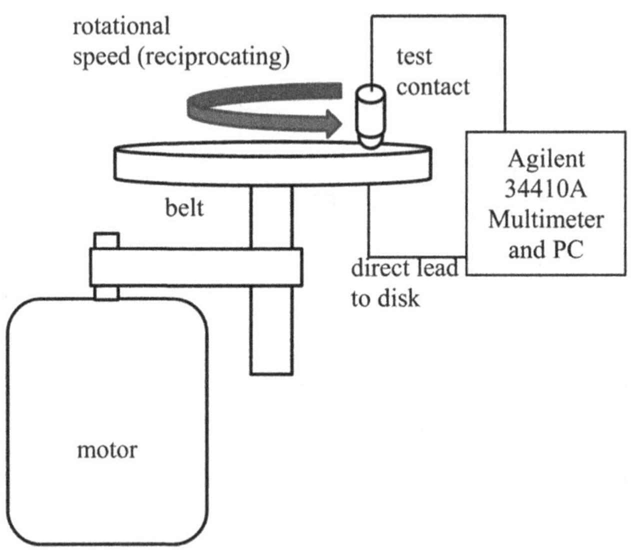

FIGURE 11.4 Schematic of a reciprocating electrical pin-on-disk test rig used in electrical connector lubricant evaluation. (Used with permission from [35].)

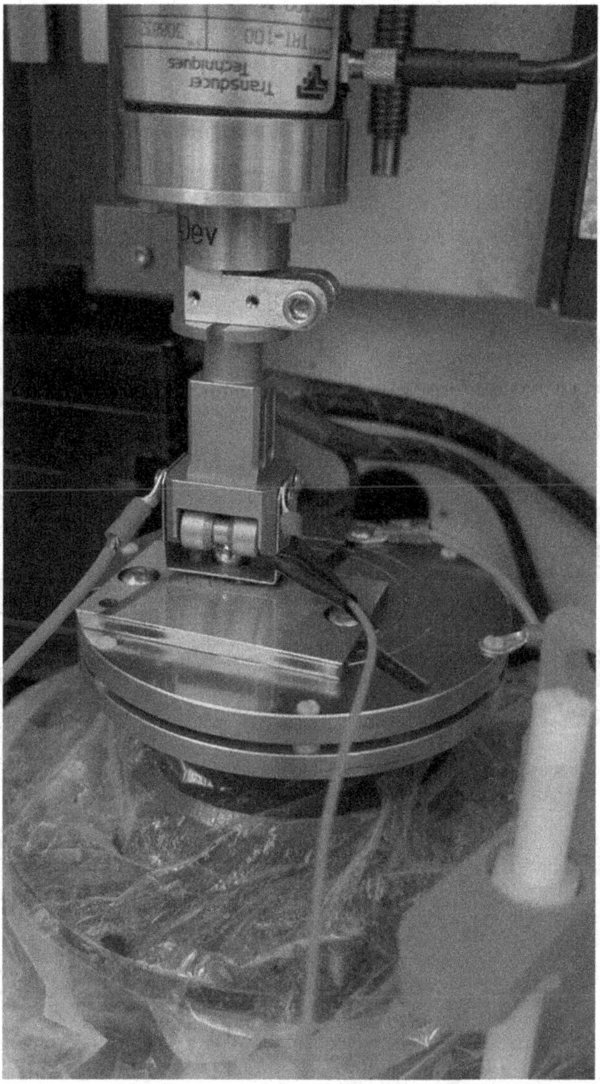

FIGURE 11.5 Image of a reciprocating electrical pin-on-disk test rig with bolted connections used in electrically conductive lubricant evaluation.

test contact, facilitating the assessment of pin-on-disk contacts. This configuration allows for precise contact resistance measurement by isolating the contact point from other extraneous resistances in the circuit. Such accuracy is crucial for understanding the actual electrical behavior of the contact under test conditions.

In the authors' most recent work [36], a spring-loaded rolling electrical contact was utilized to apply an electrical load to the center of the flat test sample, as illustrated in Figure 11.7. To securely hold the electrical contact vertically against the sample, a custom fixture was 3D-printed from nonconductive polylactic acid (PLA)

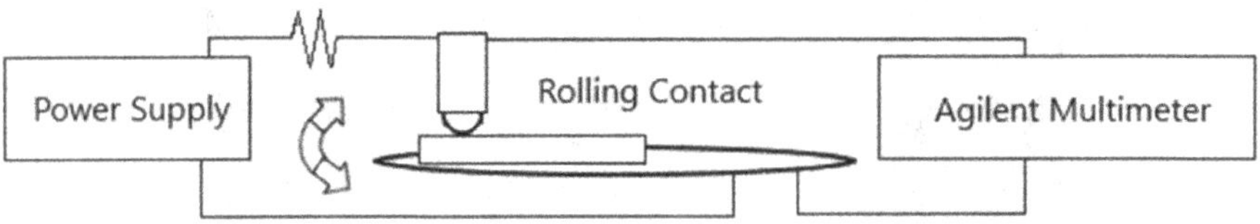

FIGURE 11.6 Schematic of a reciprocating rolling electrically loaded pin-on-disk test rig to test rolling element bearings lubricated by grease. (Electrical circuit used with permission from [24].)

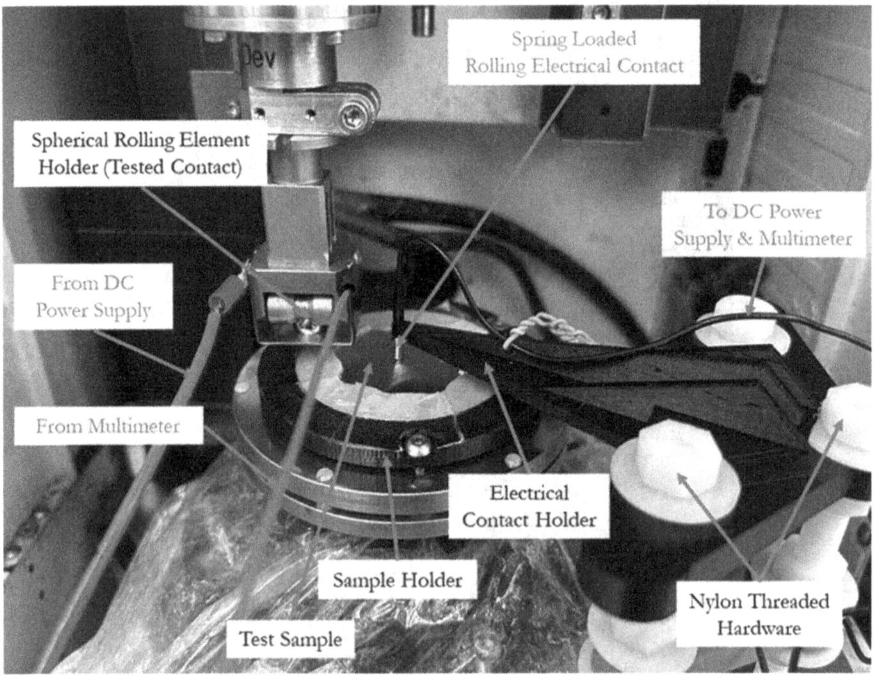

FIGURE 11.7 Schematic of a rolling electrically loaded pin-on-disk test rig to test rolling element bearings lubricated by grease [37].

filament. This holder, inspired by the design of a record player and needle, features vertical adjustability with nylon hardware to ensure the proper compression of the spring-loaded element.

The specific contact in our setup is unique, with a spherical head capable of 360° of free rotation. This design significantly reduces abrasive wear by rolling rather than sliding across the sample's surface. Placing the contact at the center of the sample minimizes the amount of movement required as the disk rotates, thereby enhancing the test's efficiency. This system is versatile, accommodating rotating and reciprocating tests while electrically insulating the setup from the Tribolab. The sample is secured to the Tribolab's universal holder with a low-profile compression ring. This

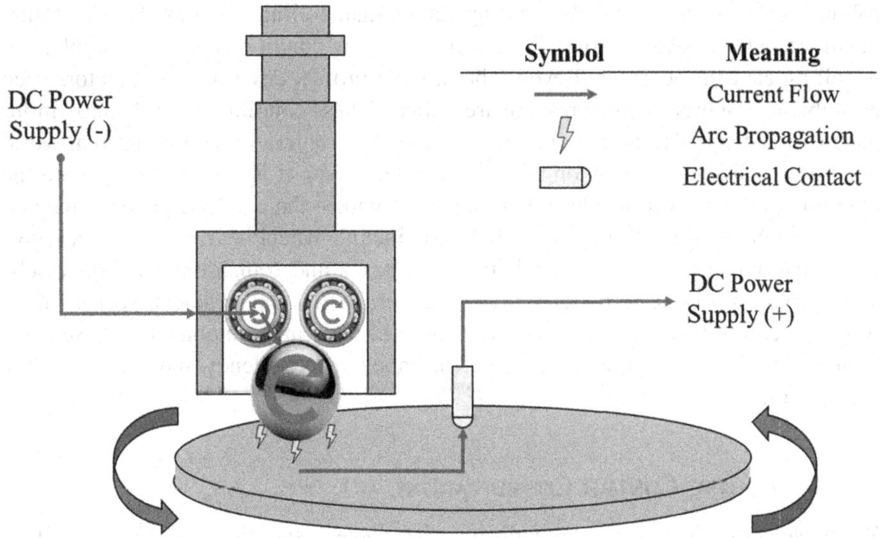

FIGURE 11.8 Current flow through element holder, roller bearings, rolling element, test sample, and electrical contact [38].

compression ring maximizes clearance with the rolling element holder and optimizes the usable surface area for testing. Another critical element of this test rig is the rolling element holder, specifically designed to accommodate spherical elements with a diameter of 9.525 mm (0.375 in.). This fixture features two rollers that support the applied load while enabling continuous rolling motion. Stainless steel ball bearings are pressed into each side of each roller to reduce friction and provide additional load support. During test operation, the fixture was continuously checked for sliding. In the event of significant abrasive damage accumulation, rolling motion is expeted to gradually decrease in favor of sliding. However, little sliding was evident throughout all tests as confirmed by SEM image analysis. SEM confirmation is reassuring considering that a major indicator of electrical and non-electrical bearing damage is increased noise, which was another audible check during test operation.

As depicted in Figure 11.7, bolted electrical connections are made to the stationary fixture's body. The leads of the power supply and multimeter are equipped with electrical eyelet connectors, which are fastened into threaded holes on the body using small screws. Electricity is conducted through the fixture's body, passing through the ball bearings and rollers, then through the spherical rolling element, and finally discharged across the grease-lubricated contact. This path is illustrated in Figure 11.8.

11.3 TEST METHODOLOGY AND PROCEDURES

The previous section introduced various moving profiles and test rig options for grease-lubricated pin-on-disk testing. While ASTM G99 provides test standards for pin-on-disk tests, it allows adjustments to the testing conditions. Typically, tests following this standard involve sliding rather than rolling motions. While sliding conditions are suitable for specific scenarios, rolling conditions are more applicable to rolling element bearings. Beyond the moving profile, environmental factors such as ambient temperature and pressure are critical. These conditions significantly influence tribological parameters and are essential for accurately simulating real-world conditions. Given the focus on electric vehicles (EVs), it is crucial to replicate the operational conditions of relevant components within the car. The primary mechanisms of interest for rolling pin-on-disk tests include wheel bearings, shaft bearings in electric motors, and gear sets within gearboxes and transmissions. Conversely, sliding pin-on-disk tests better represent general frictional contacts, such as door hinges, steering mechanisms, seat rails, and other moving components. These components are crucial for the vehicle's performance and efficiency and serve as a key area of optimization.

11.3.1 Rolling Contact Considerations

When selecting mechanical conditions for these tests, the size of the rolling element significantly influences the forces experienced at the tribological interface. Bearings in drivetrain systems vary widely in size depending on the application, and the rolling element holder used in benchtop tests often differs from the rolling elements in actual vehicle bearings. To maintain accuracy, the applied load must be carefully varied. Choosing the applied load based on the Hertz pressure and Hertz contact. Choosing an applied load based on the hertz pressure, the maximum pressure in an elastic contact with a specified curvature, offers a more accurate representation of the scaled tribological contact considers the rolling element curvature (for a sphere, this is the radius). This approach ensures that the influence of the rolling element's shape is correctly modeled. When modeling full bearings with single elements, conservatively scaling the entire bearing load to a single rolling element is critical. The load of a bearing is distributed across multiple dynamic rolling elements, so applying the whole load to a single element would result in an overestimation. Pin-on-disk tests are statically indeterminate, meaning the results depend on surface deflection, not just load and geometry. Additionally, many bearings experience multi-directional loads, including axial, radial, and moment loads. This complexity makes it challenging to determine a single applied load that fully represents bearing conditions. While the static load value on the bearing might be available from the vehicle manufacturer, dynamic conditions often require experimental validation. By carefully considering these factors, we can ensure that the test conditions accurately replicate the real-world conditions of the components in electric vehicles, thereby providing reliable and relevant data for their performance.

The temperature of the pin-on-disk test significantly impacts the results. Ambient temperature can alter the lubricant's viscosity (film thickness) and contribute to tribofilms forming on the contacting surfaces. An insulating hood or an external heating element, such as a heat lamp, can be used for precise temperature control to match the expected operating temperature in an electric vehicle. Another crucial factor to consider is the applied electric load. In EVs, leakage currents across bearings and gears can be complex, often comprising pulses and square waves, as illustrated

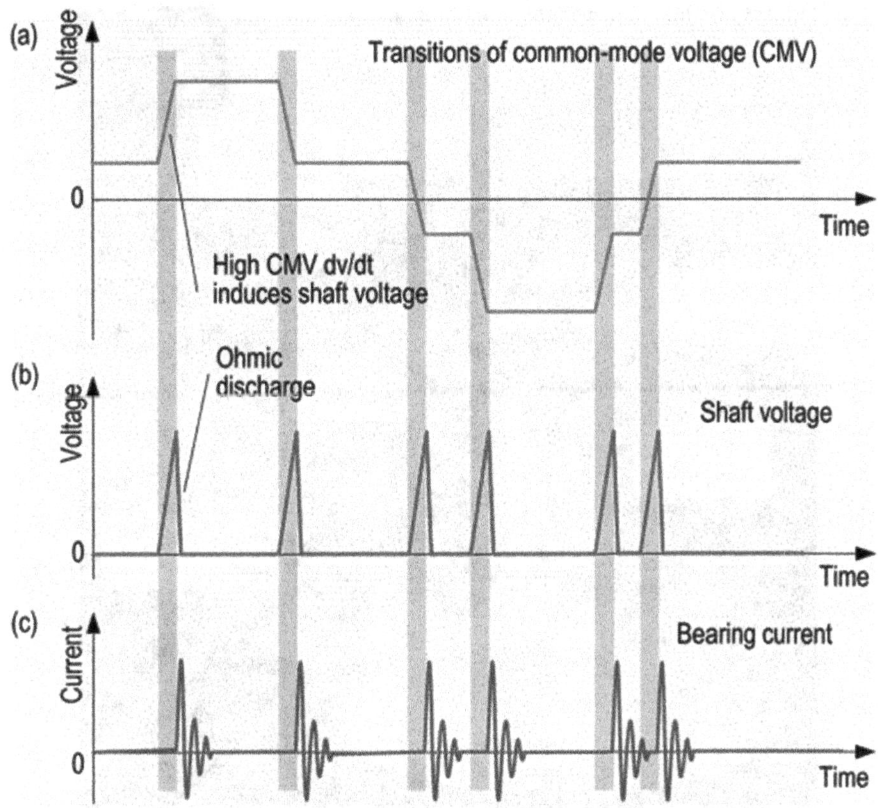

FIGURE 11.9 Example of leakage current through an EV bearing: (a) common mode voltage signal (b) shaft voltage accumulation (c) current discharge. (Used with permission from [39].)

in Figure 11.9. Depending on the leakage source and potential, the signal might resemble a simple DC or an AC (sinusoidal) current. A more idealized DC or AC electric load can benefit specific research and performance evaluations. Meticulously controlling these variables ensures test conditions accurately replicate real-world scenarios. Changing these parameters may also highlight different damage patterns associated with varying power supplies.

11.3.2 Sliding Contact Considerations

If sliding pin-on-disk contacts are tested, the results are incomparable to rolling test results. However, these tests are essential for measuring friction and wear rate for electrical connectors. The complex EV electrical systems require a vast network of electrical connectors to support the charging and electronic systems. Previous work has been conducted with various lubricants used for electrical connectors using pin-on-disk tests [33,35, 40–42]. When testing electrical connectors, the applied load is generally lower than in rolling conditions, typically ranging from 0.001 to 10N. As shown in Figure 11.10, the plated connector material is generally fixed in the

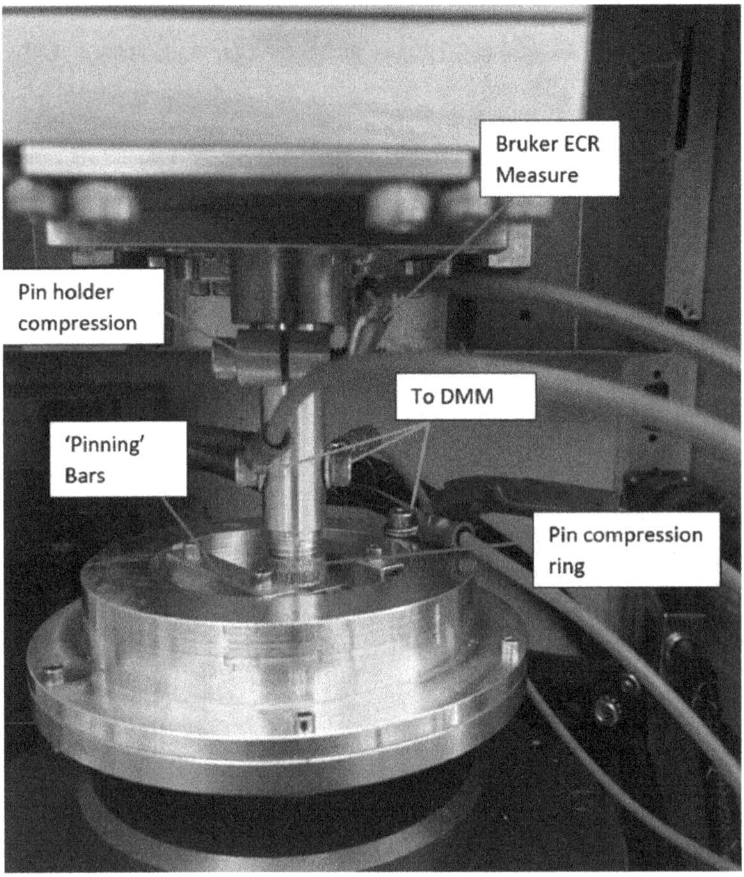

FIGURE 11.10 Test rig used in sliding electrical connector lubricant evaluation. (Used with permission from [42].)

pin-on-disk rig to prevent lateral movement during sliding tests. The plating material is usually non-ferrous metals like tin, silver, or gold. Electrical connector tests often use a reciprocating motion to simulate repeated connections and disconnections. These tests also allow replicating fretting or vibratory wear using a short stroke or high reciprocation frequency. Various sliding contacts are used in EVs, including slip rings and brushed connectors. Understanding the behavior of different electrical connectors is vital in determining the durability and robustness of EV electrical systems. It is also important to note that elements in rolling element bearings predominately rotate but may slide under certain conditions.

Key measurable metrics for sliding electrified contacts is the coefficient of friction of the system, resulting wear scar dimensions, and change in contact resistance over test duration, as supported by Jackson et al. [35].

Sliding pin-on-disk tests have also effectively assessed wear induced by electrified contact in gears for electric vehicles. Although pitting damage is less apparent than in rolling pin-on-disk tests, adding an electrical load significantly increases surface damage. The electrical load also appears to leave carbonized or degraded material on the surface due to higher induced temperatures [43]. Ionic fluids have been tested in copper pin-on-disk tests for use in electrical connectors. In these tests, a higher electric load caused arcing and surface damage. Despite these tests being aimed at electrical connectors or slip rings, the results show similarities to what is expected for EVs [44]. Evaluating the impact of lubricated sliding pin-on-disk tests provides a different approach to modeling the effects of system electrification on various machine elements and tribological interfaces. This method offers insight into how electrical loads influence wear and damage in electrified systems.

11.4 CHALLENGES AND LIMITATIONS

The variety of pin-on-disk testing in this chapter serves as a baseline for evaluating the fundamental mechanisms and critical conditions contributing to EIBD. Often, idealized conditions are examined to enable reasonable benchtop testing. These conditions present challenges and limitations when comparing experimental results to complex environments found in EVs. Many tests incorporate an AC or DC power supply, but a simple power supply might only partially capture the complexity of electrical loads EV bearings may expect. Likewise, the transient electrical conditions from deceleration and acceleration will divert different power amounts through the bearings. With this simplified electrical system, the thermal effects of varying interface power within the tribological contact might not be accurately replicated. Additionally, tests were conducted under atmospheric pressure and temperature, which may be unrepresentative of sealed electric motors.

In addition to power supply differences, testing single rolling elements may negate any dynamic variance introduced by the multibody nature of complete bearings. Single-element loading may result in comparable contact pressures to those seen in EVs. However, the complex and transient loading and load distribution in multi-element bearings may lead to different wear patterns and failure modes demonstrated in these tests. The single-element tests may only partially represent the range of loads, speeds, and temperatures experienced in real EV applications. In the simplistic and short duration of testing, fretting corrosion from prolonged exposure to vibrations and environmental conditions was also not observed.

The outlined test methodology offers significant advantages despite the challenges and limitations. Compared to complete bearing tests, single-element testing on a flat surface provides superior sensitivity in identifying early-stage damage propagation within a much shorter test duration. The flat geometry of the test disks also facilitates easier SEM image capture compared to curved samples. Furthermore, this simplified testing approach ensures consistent and repeatable results, minimizing the effects of vibrational and transient noise during data collection.

11.5 CONCLUSIONS

This chapter highlights the significance and methodology of electrified pin-on-disk testing, emphasizing its relevance to the reliability and efficiency of EV components.

Electrified conditions, such as those in EVs and other electrified machinery, pose unique challenges to powertrain components, necessitating the development of specialized lubrication strategies to mitigate wear and damage caused by electrical currents. The introduced research underscores the critical role of grease lubrication in maintaining the functionality of bearings and other mechanical elements under electrified conditions. The need for optimized greases in EVs becomes evident, and a significant portion of electric motor failures are attributed to lubrication issues.

Electrified testing using rolling pin-on-disk setups provides valuable insights into EIBD mechanisms. By isolating single rolling elements and subjecting them to controlled electrical and mechanical conditions, researchers can better understand the formation and progression of damage, namely pitting and frosting. The study also highlights the limitations of traditional testing methods, which may only partially capture real-world applications' dynamic and complex environments.

The research further explores the different testing conditions, distinguishing between rolling and sliding contacts. Rolling conditions are particularly relevant for components like wheel and shaft bearings in EVs, where the continuous rolling motion under electric load can simulate real-world wear patterns and damage mechanisms. Sliding conditions, on the other hand, are crucial for understanding the performance of electrical connectors and other sliding contacts within the EV's complex electrical system. Both testing conditions provide complementary insights, helping to create a comprehensive understanding of how electrified environments impact various mechanical components. This work also addresses the challenges of current delivery methods, such as using spring-loaded or rolling contacts for continuous electrical conduction during testing. These methods must balance maintaining a stable electrical connection while minimizing additional noise and wear, which can affect the accuracy of the test results. The innovative use of different electrical collectors, such as slip rings and brushes, also contributes to more reliable and accurate measurements.

While standardized testing for ICE lubricants is well-established, developing similar standards for EV-specific fluids is still in its infancy. This research significantly contributes to the ongoing efforts to develop reliable testing protocols that simulate realistic operating conditions, including electrical current delivery, thermal effects, and material compatibility. In conclusion, electrified rolling pin-on-disk testing with greases offers a promising approach to evaluating the impact of electrical loads on the wear and performance of lubricated mechanical components. As the complexity and prevalence of electrified systems in EVs continue to grow, such research will be crucial in guiding the development of effective lubrication solutions, ultimately enhancing electric vehicles' safety, efficiency, and reliability.

ACKNOWLEDGMENTS

This research was made possible by the generous support of the 2023 National Lubricating Grease Institute (NLGI) Academic Outreach Research Grant.

REFERENCES

1. Liu, Z., & Zhang, L. (2020). A review of failure modes, condition monitoring and fault diagnosis methods for large-scale wind turbine bearings. *Measurement*, 149, 107002. https://doi.org/10.1016/j.measurement.2019.107002.
2. Ahmed Abdalglil Mustafa, W., Dassenoy, F., Sarno, M., & Senatore, A. (2021). A review on potentials and challenges of nanolubricants as promising lubricants for electric vehicles. *Lubrication Science*, 34(1), 1–29. https://doi.org/10.1002/ls.1568.
3. Soltannia, B., Martin-Alarcon, L., Uhryn, J., Govedarica, A., Egberts, P., & Trifkovic, M. (2023). Enhanced rheological and tribological properties of nanoenhanced greases by tuning interparticle contacts. *Journal of Colloid and Interface Science*, 645, 560–569. https://doi.org/10.1016/j.jcis.2023.04.124.
4. He, F., Xie, G., & Luo, J. (2020). Electrical bearing failures in electric vehicles. *Friction*, 8(1), 4–28. https://doi.org/10.1007/s40544-019-0356-5.
5. Farooq, M. U. (2023). *Tribological Characterization of a Ball Bearing Subjected to an Electric Field : Electric Drivetrain Tribology*. Luleå University of Technology, Luleå, Sweden.
6. Walther, H. C., & Holub, R. A. (September 2014). Lubrication of electric motors as defined by IEEE standard 841–2009, shortcomings and potential improvement opportunities. In: *2014 IEEE Petroleum and Chemical Industry Technical Conference (PCIC 2014)*, San Francisco, California, USA, pp. 91–98. https://doi.org/10.10.1109/PCICon.2014.6961872.
7. Minami, I. (2017). Molecular science of lubricant additives. *Applied Sciences*, 7(5), 445. https://doi.org/10.3390/app7050445
8. Lugt, P. M. (2016). Modern advancements in lubricating grease technology. *Tribology International*, 97, 467–477. https://doi.org/10.1016/j.triboint.2016.01.045.
9. ASTM G99-23. (2023). *Standard Test Method for Wear and Friction Testing with a Pin-on-Disk or Ball-on-Disk Apparatus*. ASTM International, West Conshohocken, PA. https://www.astm.org/g0099-23.html (accessed November 11, 2024).
10. ASTM G182-13. (2018). *Standard Test Method for Determination of the Breakaway Friction Characteristics of Rolling Element Bearings*. ASTM International, West Conshohocken, PA. https://www.astm.org/g0182-13r18.html (accessed November 11, 2024).
11. Kudelina, K., Asad, B., Vaimann, T., Belahcen, A., Rassõlkin, A., Kallaste, A., & Lukichev, D. V. (2020). Bearing fault analysis of BLDC motor for electric scooter application. *Designs*, 4(4), 42. https://doi.org/10.3390/designs4040042.
12. Amjad, S., Neelakrishnan, S., & Rudramoorthy, R. (2010). Review of design considerations and technological challenges for successful development and deployment of plug-in hybrid electric vehicles. *Renewable and Sustainable Energy Reviews*, 14(3), 1104–1110. https://doi.org/10.1016/j.rser.2009.11.001.
13. Rind, S., Ren, Y., & Jiang, L. (2014). Traction motors and speed estimation techniques for sensorless control of electric vehicles: A review. In: *2014 49th International Universities Power Engineering Conference (UPEC)*, Cluj-Napoca, Romania, pp. 1–6. https://doi.org/10.1109/upec.2014.6934646.
14. Shami, U. T., & Akagi, H. (2010). Identification and discussion of the origin of a shaft end-to-end voltage in an inverter-driven motor. *IEEE Transactions on Power Electronics*, 25(6), 1615–1625. https://doi.org/10.1109/tpel.2009.2039582
15. Fiser, R., & Ferkolj, S. (1998). Magnetic field analysis of induction motor with rotor faults. *COMPEL: The International Journal for Computation and Mathematics in Electrical and Electronic Engineering*, 17(2), 206–211. https://doi.org/10.1108/03321649810203143.
16. Asakura, Y., & Akagi, H. (2013). Mechanism of shaft end-to-end voltage generation by asymmetry in an inverter-driven motor. *Electrical Engineering in Japan*, 185(1), 48–56. https://doi.org/10.1002/eej.22415

17. Prashad, H. (1994). Investigation of damaged rolling-element bearings and deterioration of lubricants under the influence of electric fields. *Wear*, 176(2), 151–161. https://doi.org/10.1016/0043-1648(94)90143-0.
18. Prashad, H. (1990). Theoretical Analysis of the Effects of Instantaneous Charge Leakage on Roller Tracks of Roller Bearings Lubricated With High Resistivity Lubricants Under the Influence of Electric Current. *ASME. J. Tribol*, 112(1), 37–43. https://doi.org/10.1115/1.2920228
19. Prashad, H. (1992). Theoretical Analysis of Capacitive Effect of Roller Bearings on Repeated Starts and Stops of a Machine Operating Under the Influence of Shaft Voltages. *ASME. J. Tribol*, 114(4), 818–822. https://doi.org/10.1115/1.2920954
20. Werner, M., Meinel, A., Graf, S., & Koch, O. (2025). A Procedure to Isolate the Electrical Capacitance of Tooth Contacts From Charging Curve Measurements of a Gear Stage. ASME. J. Tribol, 147(5), 051113. https://doi.org/10.1115/1.4067975
21. Jackson, R. L., Saha, S., & Janik, J. R. (2025). A Statistical Prediction of Electrical Discharge Initiation and Semi-Analytical Transient Mixed Lubrication Model of a Rolling Element. *ASME. J. Tribol*, 147(5), 051103. https://doi.org/10.1115/1.4066520
22. Balakrishna, G. V., & Gnanamoorthy, R. (2025). Current–Voltage–Friction Characteristics of Grease in Electromechanically Loaded Sliding Contacts. *ASME. J. Tribol*, 147(5), 051105. https://doi.org/10.1115/1.4066758
23. Notay, R. S. (2025). A Brief Review on the Tribological Effects of Electrically Induced Bearing Damage. *ASME. J. Tribol*, 147(5), 051112. https://doi.org/10.1115/1.4067831
24. Bond, S., Jackson, R. L., & Mills, G. (2024). The influence of various grease compositions and silver nanoparticle additives on electrically induced rolling-element bearing damage. *Friction*, 12(4), 796–811. https://doi.org/10.1007/s40544-023-0837-4
25. Chatterton, S., Pennacchi, P., & Vania, A. (2016). Electrical pitting of tilting-pad thrust bearings: Modelling and experimental evidence. *Tribology International*, 103, 475–486. https://doi.org/10.1016/j.triboint.2016.08.003.
26. Janik, J. R., Saha, S., & Jackson, R. L. (2024). Rolling element damage and the influence of test conditions and conductive nanoparticle grease additives. In: *Presented at the 32nd ICEC and 69th IEEE Holm Conference on Electrical Contacts*, Annapolis, MD, USA, 2024.
27. SWRI. (2023). *Southwest Research Institute Develops Device to Test Friction, Wear Associated with EV Fluids.* Southwest Research Institute. https://www.swri.org/press-release/southwest-research-institute-develops-device-test-friction-wear-associated-ev-fluids.
28. Suzumura, J. (2016). Revention of electrical pitting on rolling bearings by electrically conductive grease. *Quarterly Report of RTRI*, 57(1), 42–47. https://doi.org/10.2219/rtriqr.57.1_42.
29. Hobelsberger, M. P., & Zlatimir, Z. (2009). Method for monitoring the shaft current and or the insulation, Patent Application, PCT/EP2008/066872, 2009.
30. Noguchi, S., Kakinuma, S., & Kanada, T. (2010). Measurement of direct current voltage causing electrical pitting. *Journal of Advanced Mechanical Design, Systems, and Manufacturing*, 4(6), 1084–1094. https://doi.org/10.1299/jamdsm.4.1084.
31. Demas, N. G., Lorenzo-Martin, C., Luna, R., Erck, R. A., & Greco, A. C. (2023). The effect of current and lambda on white-etch-crack failures. *Tribology International*, 189, 108951. https://doi.org/10.1016/j.triboint.2023.108951.
32. Guo, L., Mol, H., Nijdam, T., de Vries, L., & Bongaerts, J. (2023). Study on the electric discharge behaviour of a single contact in EV motor bearings. *Tribology International*, 187, 108743. https://doi.org/10.1016/j.triboint.2023.108743.
33. Tajedini, M., Arole, K., Ghasemi, Z., Azhdari, R., Green, M. J., & Liang, H. (2024). Tribo-electrical evaluation of conductive fluid film of Ti_3c2tz mxene-containing lubricant. *Materials Advances*, 5(12), 5063–5069. https://doi.org/10.1039/d3ma01036h.

34. Gohar, R. (1988). *Elastohydrodynamics*. Horwood Limited, Ellis, p. 319.
35. Jackson, R. L., Coker, A. B., Tucker, Z., Hossain, M. S., & Mills, G. (2019). An investigation of silver-nanoparticle-laden lubricants for electrical contacts. *IEEE Transactions on Components, Packaging and Manufacturing Technology*, 9(2), 193–200. https://doi.org/10.1109/tcpmt.2018.2887402.
36. Janik, J. R., Saha, S., Jackson, R. L., & Mills, G. (2024). Exploring the boundaries of electrically induced bearing damage in grease-lubricated rolling contacts. *Lubricants*, 12(8), 268. https://doi.org/10.3390/lubricants12080268.
37. Janik, J. R., Saha, S., Jackson, R. L., & Mills, G. (2025). Rheological and electropitting performance of electric vehicle motor greases with various nanoparticle greases. *Journal of Tribology*, 147(5). https://doi.org/10.1115/1.4067489.
38. Janik, J. R. (2024). *Exploring Electrically Conductive Nanoparticle Additives in Greases for Electric Vehicle Applications*. Master of Science, Auburn University, Auburn, Albama, USA.
39. Hadden, T., et al. (2016). A review of shaft voltages and bearing currents in EV and HEV motors. In: *IECON 2016 – 42nd Annual Conference of the IEEE Industrial Electronics Society*, Florence, Italy, IEEE, pp. 1578–1583. https://doi.org/10.1109/IECON.2016.7793357.
40. Kaneko, S., Taura, H., Fukasawa, R., & Kanai, H. (2015). Lubrication characteristics of electric sliding contacts consisting of rotating circular grooved disk and stationary rider with spherical surface under lubricated condition. *Journal of Tribology*, 138(1). https://doi.org/10.1115/1.4031393.
41. Mao, F., Wiklund, U., Andersson, A. M., & Jansson, U. (2015). Graphene as a lubricant on Ag for electrical contact applications. *Journal of Materials Science*, 50(19), 6518–6525. https://doi.org/10.1007/s10853-015-9212-9.
42. Crilly, L., Jackson, R. L., Mills, G., Bond, S., & Bhargava, S. (2022). An exploration of the friction, wear, and electrical effects of nanoparticle enhanced and conventional lubricants. *IEEE Transactions on Components, Packaging and Manufacturing Technology*, 12(11), 1757–1770. https://doi.org/10.1109/tcpmt.2022.3225069.
43. Cao Romero, J. A., Reyes-Avendaño, J., Soriano, J., Farfan-Cabrera, L., & Erdemir, A. (2022). *A Pin-On-Disc Study on the Electrified Sliding Wear of EVs Powertrain Gears*. SAE Tech. Paper 2022-01-0320. SAE International, Warrendale, PA. https://doi.org/10.4271/2022-01-0320.
44. Wen, X., Yuwen, F., Ding, Z., Zhang, W., Yao, R., & Lu, J. (2019). Electric arc-induced damage on electroless Ag film using ionic liquid as a lubricant under sliding electrical contact. *Tribology International*, 135, 269–276. https://doi.org/10.1016/j.triboint.2019.03.019.

Index

Note: **Bold** page numbers refer to tables and *italic* page numbers refer to figures.

F

Printed in the United States
by Baker & Taylor Publisher Services